Prologue

When the First Edition of Overcurrents and Undercurrents was originally published in 1996, the sub-title was: All about GFCIs. The focus of the text was on how the principle of the residual current device, plus electronics, added a whole new dimension to electrical safety.

Mention was made in the original text that in the future, new electronic devices as yet undefined, would evolve to add to electrical safety. During that same time frame of the mid- 1990s, several manufacturers were busily engaged in their R & D facilities with the next generation of electrical safety devices, which have become known as Arc-Fault Circuit-Interrupters, or AFCIs. These devices are now on the market. They operate on an entirely different principle from the residual current devices such as GFCIs and GFP.

The Second Edition of Overcurrents and Undercurrents has been revised to include and to describe this new AFCI technology, and the sub-title has been changed to: All about GFCIs and AFCIs.

Chapter 11 has been revised to include changes relative to GFCIs, GFP, and AFCIs in the 1999 edition of the National Electrical Code.

The people at Cutler- Hammer Division of the Eaton Corp., a major source of this new technology, have most graciously helped me to include Cutler- Hammer AFCI information in the book.

The author wishes to extend a special word of thanks to John Wafer, Clive Kimblin, Robert Clarey, and their staffs at Cutler-Hammer for their help in understanding and describing this new AFCI technology.

7-4-00
E. W. Roberts

To Prof. Charles F. Dalziel,
of the University of California at Berkeley,
who first introduced the GFCI concept
into the USA to protect people
from lethal electrical shocks.

Published by :

Reptec
Eight Brandon Ln
Mystic, Ct 06355 - 3103
Phone - (860) 536-4496
FAX - (860) 536-4496
email: Reptec1@aol.com

Printed by :

Copy Cats
458 Williams Street
New London, CT 06320
Phone - (860) 442-8424
FAX - (860) 437-8697

First Edition - 1996
Second Edition - 2000

ISBN 0-9674323-1-6

Table of Contents

Overcurrents and Undercurrents
Electrical Safety Advances Through Electronics

How the GFCI and related offspring are creating a revolution in electrical shock and electrical fire protection.

TN-C = Common Grounded System
TN-S = Separate Grounded System
TN-C-S = Part Common and Part Separate Grounded System
TT = Ground Connections at Each Utilization Equipment.
Earth the only common conductor.

Characteristics. The variations - Self- Contained Type, Circuit Breaker Type, Receptacle Type, Portable Type, Power Supply Cord Type, and Plug Type.

Index of Illustrations

Figure	Title	Page

Introduction

A revolution is underway in electrical wiring safety of buildings.

The application of electronics to electrical protective devices has greatly accelerated the normal evolutionary process towards our ultimate building electrical safety goal: electrical systems which are electrical shock–free and electrical fire–free.

In this text, we plan to review the history of these changes, to discuss the present state–of–the–art, and to dream a bit of what may lie ahead.

Thomas Edison invented the first protective device–the fuse–a controllable weak link in a wire. A fuse at a safe and convenient location would melt at a predetermined value of current, opening the circuit and protecting the remainder of the circuit from damage. After the cause of the trouble had been eliminated, the fuse link could be replaced, restoring power. The fuse has experienced many improvements since, but the fuse principle is the same. Sometime later came the circuit breaker, which could be reset after it tripped, eliminating the need to replace a fuse link.

For decades after the introduction of the fuse and the circuit breaker, attention has focused on improving "overcurrent" devices and attempting to insure through adequate grounding that wiring faults to ground became "overcurrents," thus activating the overcurrent device.

The mindset of many in the electrical industry was not unlike that of the railroad industry people at the turn of the century. The railroad people viewed their role as "running trains" rather than as "transporting people and freight efficiently and economically."

Likewise, certain electrical industry people focused on grounding and electrical overcurrent protection hardware as THE THING.

Of course, THE THING is protecting people and the electrical system from any and all undesirable currents, be they overloads, short circuits, high level ground faults, low level ground faults or arcing faults. Overcurrent protection cannot protect against harmful "undercurrents"– low level ground faults and arcing faults.

Low level faults are the ones that are killing people through electrical shock and that are starting electrical fires through arcing ground faults and arcing faults.

Just about all the buildings in the USA were constructed in conformance with some issue of the National Electrical Code, which most assuredly required overcurrent protection and grounding. Yet even now, several hundred people are killed or seriously injured from electrical shock annually in or around buildings at voltages below 600 v. Hundreds of others are killed or injured annually from building fires caused by electricity.

Building electrical fire damage runs into the hundreds of millions of dollars annually, according to the statistics of the National Fire Protection Association. Electrical equipment arcing represents the No. 1 cause of fires in stores and manufacturing properties.

The advent of electronics has made it possible to come closer to total electrical system protection , including protection against low level ground faults and arcing faults.

This text is about how the principle of the Residual Current Device (RCD), as applied in the USA by the Ground Fault Circuit Interrupter (GFCI) and Ground Fault Protection of equipment (GFP), and related electronic creatures, some of which use a different principle than the RCD, such as the Arc Fault Circuit Interrupter (AFCI), are revolutionizing electrical building safety.

Chapter 1

The Beginning of
Electric Shock Concerns

Ben Franklin

The first concerns for the safety of people and property exposed to electricity coincides with the first exposure of society to electricity.

We will begin our story with mention of certain relevant developments and inventions by one of the greatest "Jacks of all trades and masters of many" ever produced in the USA, Benjamin Franklin.

Ben was born on Jan. 17, 1706, in Boston, Mass., the 15th in a family of 17 children. Because of his father's need for his services in the family candle making shop, he received only 2 years of formal schooling.

By his own initiative he became a highly educated man. At 17 he ran away to Philadelphia, ultimately becoming a printer, publisher, civic leader, scientist, inventor, statesman, and patriot, but,we will touch only on his electrical scientific and inventive activities here.

His most famous electrical experiment was conducted in Philadelphia in 1752, when he flew a kite in a thunderstorm and proved that lightning is actually electricity. The spark from the key attached to the kite string proved that the thunderstorm cloud was electrostatically charged. We know today how lucky he was to survive the experiment, as a lightning strike to the kite string might have ended the career of Dr. Franklin.

Franklin went on to invent the lightning rod, and promote its use. One of his lightning rod systems was installed on his own home and it was struck by lightning, resulting in no damage.

The system became a financial success, and is essentially the basic system used today.

Ben was also one of our earliest recorded nonfatal electrical shock victims. One day he was experimenting with trying to kill a turkey by electrical shock. The experiment went awry, and Ben received the shock instead of the bird. He was knocked unconcious. Upon recovering he was quoted as saying "I meant to kill a turkey, and instead, I nearly killed a goose."

Thomas A. Edison

The story of the safe use of electricity as we know it today really began with the work of Thomas A. Edison , certainly a great inventor if not the greatest in history.

Edison was born on Feb. 11, 1847, in Milan, Ohio, the youngest of 7 children. As with Franklin, many biographies exist on his life, and we shall touch only briefly on a few subjects relevant to our story.

Of his many basic inventions, the ones in 1879 of the incandescent lamp and a complete electrical system to supply the lamp are of primary significance. Electric lights existed before Edison's invention of a practical incandescent lamp, but they were expensive, ponderous, high maintenance, carbon–arc lamps, used for street lighting.

Edison believed that if current were passed through a conducting filament with enough electrical resistance to cause it to heat up to an incandescent temperature, as long as no oxygen were present to allow the material to burn, the material would continue to glow, giving off light.

He employed glass blowers to make the lamp envelope and developed vacuum techniques to remove the oxygen from the interior of the glass envelope.

His biggest challenge was finding the right material for the filament. He tried hundreds of materials without success. Most of us would have given up. He never considered the unsuccessful tests as failures. In his mind, the tests were successful, because they proved that those particular materials would not work. Finally, he tried bamboo, which had been converted to a form of graphite. It worked, and the rest is history.

Later, tungsten became the standard filament material. I have an exact replica of Edison's original lamp, one of a batch made at GE's Nela Park to commemorate Edison's 100th anniversary. It's very rugged, and still works.

Edison realized that his lamp required a complete, safe electrical system in order for electrical lighting to compete with, and ultimately to replace gas lighting. He chose approximately 120 v DC initially, and later 120/240 v DC for the system. He designed the generator and all the necessary components. He started up the first generating station in New York in 1882. He anticipated the need to provide protection for the system, and he invented the fuse.

The fuse has gone through many changes since first invented but its purpose is the same: to create a known weak link in the electrical system which will melt or "fuse" when a current reaches a value which is so high that if allowed to continue, will cause damage to the wiring or equipment on the circuit. The fuse location could be selected for easy replacement and service.

Edison met with great opposition from the gas industry, which resorted to every half–truth and dirty trick in the book to prevent the use of the Edison electrical system. However, the inherent advantages of incandescent electric lighting over gas lighting triumphed, and the Edison system was widely adopted in major cities.

George Westinghouse

Westinghouse was born on Oct. 6, 1846, in Central Bridge, NY. Among the many significant inventions he helped to promote and to market those related to the electrical transformer and the components necessary for an AC system, versus the Edison DC system, are most important to this discussion.

For our less technically inclined readers, the big difference between AC and DC is that in an AC system, the voltage can be readily changed up or down by means of transformers. Thus the voltage from the generator is stepped up to a high voltage for transmission over long distances, and then stepped down for use by the customer.

Since electric power, or real energy, is the product of the resistive current times the voltage, doubling the voltage will double the power, with the same current. Likewise, if the transmission voltage is 1000 times the utilization voltage, it is possible to transmit 1000 times the energy through the same size conductor. The conductor size is determined by the amount of current to be carried. Of course, with the same current and 1000 times the voltage, it is possible to supply 1000 times as many customers from the same transmission line. The voltage must be lowered to a utilization value for use by customers.

The inherent advantages of AC over DC are obvious. It is interesting that Edison, then the entrenched power, used the same dirty tricks against Westinghouse to stop AC as the gas industry used against him. One attempt to stop AC was the claim that it was unsafe. Edison tried to focus all the bad press he could on AC by promoting its use for killing people in the electric chair. It is true that the danger of death is greater for a given intensity and duration of shock from AC than from DC. This will be discussed in detail in Chapter 7.

Prof. Theodore Bernstein, of the University of Wisconsin, in his Feb. 1973 and Sept. 1976 articles in the IEEE Spectrum, provides interesting insights. In 1888, Edison ran an advertisement in the newspapers, challenging Westinghouse to what might be called an "electrical duel". Westinghouse would hold onto AC and Edison to DC. They would both start off at 100 v. The voltage would be increased in 50 v increments, with Edison going first, until one person cried "uncle." Edison warned his opponent that his experiments indicated that 150 v AC could be fatal. Westinghouse ignored the challenge.

By 1886 the State of New York was looking for a better way to execute prisoners. Hanging was not doing an efficient job, probably due to improper knots or procedures. A commission was created to recommend a better way. Beheading by a guillotine was discarded because of a desire for an intact body. The garrote was eliminated as too cruel and medieval. A dentist on the commission, Dr. Alfred Southwick, noting the quick death of a Buffalo, NY man who contacted the brushes of a generator, promoted electrocution. The result was that in June 1888, NY passed a law that after Jan.1, 1889, electrocution would become the capital punishment for New York State.

Talk about being in the wrong place at the wrong time, enter William Kemmler. Kemmler murdered his girl friend with an axe in March, 1889, just in time to be convicted and become the first person legally electrocuted.

Kemmler became a pawn in the Edison–DC versus Westinghouse–AC fracas. Harold Brown, a New York "electrician/ engineer," was awarded the contract to make the electric chair and system. Brown had been doing work for Edison previously. A Westinghouse AC dynamo was selected. Brown had difficulty locating a new one thanks to the impeding efforts of Westinghouse, but he finally located a used one. Brown's electrical system was used, but his electric chair was replaced by one designed by a doctor, Dr. George Fell.

We won't dwell on the grisly details. Suffice it to say that Kemmler experienced two shocks. No readings were taken as to current and voltage. The first shock lasted 17 seconds, and the second one lasted 70 seconds. Smoke was observed after the second one. In short, " this ain't no way to kill a body!" The effects of current on the human body is covered in detail in Chapter 7.

In spite of all the negative publicity, the AC system, with it's inherent technical and economic advantages has triumphed, and has allowed us to do the many electrical and electronic miracles we'll be discussing, which would have been much more difficult with DC.

Chapter 2

Overcurrent Protection
Fuses and Circuit Breakers

In the early growth of building electrical systems, the fuse played a vital role in protecting the systems from unsafe over-currents - overloads, short circuits, and high level ground faults.

From its humble beginning in 1880 as a simple copper " safety conductor" to today's sophisticated, current- limiting device, the fuse is recognised worldwide as a reliable, cost effective overcurrent protective device.

Fuses have several inherent characteristics that distinguish them from other types of overcurrent protective devices. They are simple, reliable, have very high interrupting ratings, and have an inherent ability to respond extremely fast under short - circuit conditions. Today's modern current - limiting fuses have high interrupting ratings, up to 300,000 amperes, and can clear an overcurrent in less than one- half cycle, or 0.008 seconds. They are available in a wide range of sizes from 1/ 500th of an ampere to 6,000 amperes.

Because of their distinctive characteristics, fuses are used for many residential, commercial, and industrial purposes. They are used at services, as well as on feeders and branch circuits. They are also used as supplementary overcurrent protection in portable appliances, hot tubs, spas, and many electronic devices.

Fuses can be designed with special time- delay performance characteristics, making them particularly suited to protect motors and motor branch circuits. Dual- element, time- delay fuses can be sized to override harmless motor inrush currents and provide motor running protection as well as branch circuit, short- circuit and ground- fault protection, all with one device.

An inherent benefit of today's modern fuse is its indigenous

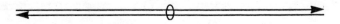

current- limiting ability. Because a fuse is a purely electrical device rather than an electromechanical device, it can respond very rapidly under short circuit conditions. The fuse's ability to be current- limiting provides the system designer with a good tool to solve complex protection problems. For example, the current- limiting properties of fuses can be used to protect other devices, such as circuit breakers with lower interrupting ratings, circuit components, and sensitive electronics.

To summarize, fuses are an efficient, cost effective way to provide overcurrent protection, particularly on motor circuits and where high levels of fault current can be anticipated. However, the need for fuses to be replaced when they perform their function, with all the inconvenience, delay, and cost of replacements, left them vulnerable to a more "user friendly" solution to overcurrent protection.

Also, the early household plug fuse design, commonly used on circuits from 15A to 30A, could easily be misapplied. The screw thread for all sizes was identical, so that a 15A fuse could be replaced with a 20, 25, or 30A fuse, allowing the circuit conductors to become overloaded. Also, a penny could easily be inserted under a blown fuse, negating any overcurrent protection and adding to the hazards.

The electrical inspectors, insurance interests, and others concerned with safety clamored for a solution to this problem. The fuse industry responded with "Type S" fuses and fuse holders, which had unique screw threads for each rating. Adapters for existing fuse holders to allow conversion to a Type "S" configuration, were made available. The adapters were designed so that once installed in an old style fuse socket, they could not be removed. The problem focused attention on the need for a type of overcurrent protection which would overcome the disadvantages of fuses.

The time had come for the circuit breaker. The circuit breaker is defined in the National Electrical Code as:

> "A device designed to open and close a circuit
> by nonautomatic means and to open the circuit

> automatically on a predetermined overcurrent
> without damage to itself when properly applied
> within its rating."

The circuit breaker can be used as a switch and can be easily reset by the user after it has tripped. In this text we shall confine our discussion to what are considered "low voltage" building distribution circuit breakers of the molded–case type, used on circuits of 600 v or less. Electric utilities and large industrials use high voltage circuit breakers, which usually receive the commands to trip from separately located relays and usually interrupt the current under oil or in a blast of air through arc chutes. Also, air circuit breakers of the open frame type are used by industry for heavy duty applications, usually at voltages of 1000 v max. and at currents in the 600 to 4000 A range.

The circuit breaker was born in the 1920's as a result of the invention by Westinghouse engineers of the "DE–ION" arc extinguisher for use initially in large oil circuit breakers. Westinghouse literature states that although too large in its initial form to be used in small circuit breakers, the arc extinguisher was eventually modified into a more compact size.

The first compact, workable circuit breaker was developed in 1923 when the modified arc extinguisher was coupled with a thermal tripping mechanism. Westinghouse states that it was not until 1927 that the first molded–case circuit breakers, capable of interrupting 5000 A at 120 v AC or DC, were intruduced to the market.

Acceptance was immediate. Today, molded–case circuit breakers dominate the residential building market and are a significant factor in commercial and industrial buildings.

Molded–Case Circuit Breaker Construction and Operation

For the purposes of this text, our discussion of " breakers" will essentially refer to molded–case circuit breakers, though many of the principles we cover also apply to other types.

The two basic operating functions of a breaker are: sensing and interrupting overcurrents. The means of sensing which a breaker uses categorizes it as being either a thermal–magnetic, dual–magnetic, or electronic type of breaker. All of these breakers are of the inverse time type, which means the higher the current the faster they trip, and the lower the current the slower they trip. Actually, for the thermal–magnetic and dual–magnetic types the inverse time function applies to the overload range, but once the current is in the high magnetic range, the trip is "instantaneous"–usually less than a half a cycle. For 60 Hertz (60 cycle per second) AC systems, this is 1/120th of a second.

The thermal–magnetic breaker uses a heat–sensitive bimetallic element, influenced by the normal current path, to detect overload currents. It uses a magnetic circuit influenced by the very high currents that occur during short circuits or high level ground faults, to detect high current faults.

A dual–magnetic breaker uses magnetic circuits for both the overload and the short circuit or high level ground fault tripping functions. In order to emulate the inverse time function, a dashpot is usually used for the overload condition.

The electronic breaker is only "electronic" for the sensing functions, and substitutes electronic circuitry for the electromechanical sensing components of the other two types. It senses overloads, short circuits, and high level ground faults electronically. The first electronic breakers had analog circuitry and the newer generations tend to be digital.

It should be kept in mind that the actual interruption of the current is accomplished essentially the same way in all the breakers–through arc chutes. Circuit breakers with "current limiting" capabilities are also gaining usage.

It is an intriguing idea to be able to interrupt the current electronically, as well as detecting the undesirable currents, making a true "electronic" circuit breaker. So far, this has not been accomplished in a practical way for circuit breaker applications. We're talking about interrupting current ratings of 10,000, 22,000 and 65,000 A for these applications. It is not an easy task.

The author was visited by a venture capitalist and an electronics engineer once to discuss a "truly electronic" circuit breaker they had developed. It relied on interrupting the current very, very early, when it was still considerably less than an ampere. It would be essential that the current be interrupted when very small, and with ultimate reliability, in order for the device to survive. They demonstrated a prototype and showed test data. I asked them how their circuitry could differentiate between a normal build-up in current when an appliance or motor was started and the beginning of a true overload or fault condition. They haven't been back. Maybe someone will solve the problem in the future.

Let's look at the basic thermal-magnetic circuit breaker to

THE SENSING "THERMAL UNIT" IS A BIMETALLIC BAR

If metal "B" 's coefficient of expansion is greater than metal "A" 's, the bar will bend upwards.

Figure 1

Courtesy: GE

HOW IT OPERATES...
OVERLOAD

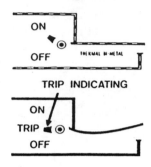

Figure 2
Courtesy: GE

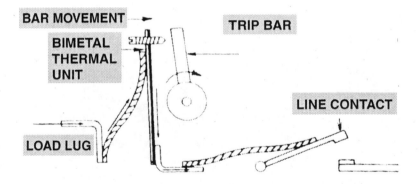

BASIC OVERLOAD SENSING MECHANISM

Figure 3
Courtesy: GE

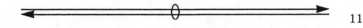

understand how it works. This type of breaker senses overloads thermally by means of a bimetallic element. Fig. 1 shows a "sandwich" made of two metals with different coefficients of expansion, which means that as they heat up, they expand at different rates. In the illustration, if metal B has a higher coefficient of expansion than metal A, the bar will bend upwards

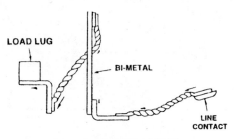

DIRECT HEATED
10-40 AMPS

Figure 4

Courtesy: GE

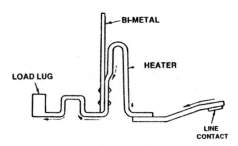

INDIRECT HEATED
50-800 AMPS

Figure 5

Courtesy: GE

when heated. Of course, the higher the current through the bar, the faster it will heat up and the sooner it will bend. Fig. 2 and Fig. 3 show how this is applied to the overload tripping mechanism of a circuit breaker.

The lower rated breakers usually have directly-heated bimetals, in which the normal current path is directly through the bimetal. Fig. 4 illustrates this, and for typical industrial breakers this design is used in the 15 to 40 A ratings. The higher ratings usually have indirectly heated bimetals, as illustrated in Fig. 5. For a typical industrial breaker, this design is used in the 50 to 800 A ratings.

Short circuit and high level ground-fault protection are achieved by using the high magnetic forces that are generated by the fault current to trip the breaker. Fig. 6 illustrates the way magnetic forces are generated by current flow. Fig. 7 shows how this force is harnessed to attract an armature to an electromagnet, tripping the breaker through a linkage. Fig. 8 shows how the value at which the magnetic trip occurs can be adjusted by varying the air gap between the armature and the magnet.

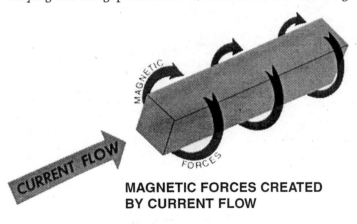

MAGNETIC FORCES CREATED BY CURRENT FLOW

Figure 6

Courtesy: GE

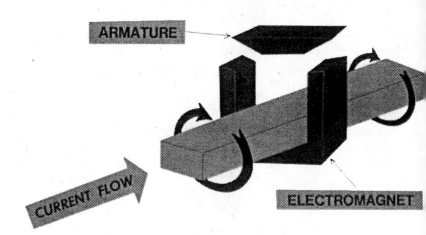

ELECTROMAGNET ENERGIZED BY MAGNETIC FORCES ATTRACTING ARMATURE. AS THE CURRENT INCREASES, THE MAGNET'S PULL INCREASES.

Figure 7

Courtesy: GE

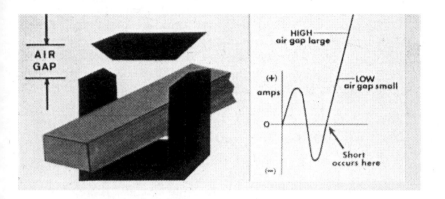

BY VARYING THE AIR GAP, THE PICK-UP POINT AT WHICH HE CIRCUIT BREAKER TRIPS CAN BE ADJUSTED.

Figure 8

Courtesy: GE

Some variants of the molded case circuit breaker are the "instantaneous only" circuit breaker, normally called the Instantaneous Circuit Breaker, and the Molded–Case Switch.

Instantaneous Circuit Breakers

If you leave out the overload sensing capability of a circuit breaker, you have an Instantaneous Circuit Breaker. These have use in motor circuits, where overload protection is provided separately. They are not listed by a third - party testing laboratory, such as Undrwriters Laboratories, Inc., to prevent their use on general purpose circuits, and are "recognized components." They become a component in a motor starter, which is usually listed as a complete unit, with separate overload protection.

Molded–Case Switches

If you leave out all the overcurrent sensing capability, you have a Molded–Case Switch. The arc chutes are still there, though, so there is an ability to interrupt high currents when the switch is turned off manually.

The molded–case circuit breaker, particularly the electronic types, set the stage for the next major advancement in electrical circuit protection–the ability to detect and to trip when low level ground faults exist.

Chapter 3
Grounding Practices in North America

The subject of "grounding" is very complex and is discussed in frequent seminars, industry magazine articles, and texts devoted to the subject. One of the classic books on grounding for those who would like to delve into more detail is "Grounding Electrical Distribution Systems for Safety," by Eustace C. Soares. The book was first published in 1966, based on the 1965 issue of the National Electrical Code.

The book is extremely helpful in explaining many of the things that grounding can and cannot do. Rights to the text have been obtained by the International Association of Electrical Inspectors, who revise the text to bring it up to the current issue of the National Electrical Code.

We shall confine our discussion to how grounding relates to the basic theme of this text.

The first point that should be made is that "grounding" must be looked at as more than the connection of one or more conductors of an electrical system to the earth through a suitable grounding electrode.

Article 250 of the National Electrical Code covers the general requirements for grounding and bonding of electrical systems. The best justification for "grounding and bonding of electrical installations" is given in the two Fine Print Notes of Section 250–1. Scope, in the 1996 NEC:

> "FPN No. 1: Systems and circuit conductors are grounded to limit voltages due to lightning, line surges, or unintentional contact with higher voltage lines, and to stabilize the voltage to ground during normal operation. Equipment grounding conductors are bonded to the system grounded conductor to pro-

vide a low impedance path for fault current that will facilitate the operation of overcurrent devices under ground fault conditions."

"FPN No. 2: Conductive materials enclosing electrical conductors or equipment, or forming part of such equipment, are grounded to limit the voltage to ground on these materials and bonded to facilitate the operation of overcurrent devices under ground fault conditions."

Most things in this world have both advantages and disadvantages, and although the advantages of an adequate grounding system for an electrical system far outweigh the disadvantages, disadvantages do exist.

Before the introduction of the Ground Fault Circuit Interrupter (GFCI) and Ground Fault Protection of equipment (GFP), which were the first devices to apply the principle of "residual current" protection in the USA, there were four basic ways to protect people and property from the harmful effects of electricity: *isolation, insulation, overcurrent protection, and grounding.*

"**Isolation**" means keeping people and objects away from the electricity. In essence, this uses distance through air as the insulator. The most obvious example of this is the use throughout the USA of bare conductors on distribution, distribution supply, and transmission lines, elevated in air above our streets and in rights-of-way. A critic might claim that the lines should have insulated conductors, but society has long accepted the tradeoff of somewhat less safety for greatly reduced cost-ultimately reflected in the cost of electricity to the consumer.

Air is one of the best and most cost effective insulators. Birds perch on one bare energized phase conductor of power lines regularly and experience no ill effects. If a ground wire were in reach of a bird at the instant it was touching a phase conductor, it would be incinerated instantaneously by a ground fault.

Power company linemen work on bare energized power lines

with bare hands on a routine basis from insulated bucket trucks. A grounded bucket would cause instant death.

Thus the successful use of bare power lines depends on isolating the line from everything, including ground. The grounded shielding conductor, often installed on transmission lines high above the phase conductors, provides an electrical "umbrella" from lightning.

"**Insulation**" refers to the actual covering of energized surfaces with insulating material which will protect any person or object which comes in contact with the outer surface of the insulated conductor from receiving a signficant flow of electricity from the conductor.

"**Overcurrent Protection**" is the sensing and interrupting of overload, short circuit, and high level ground fault currents. Overload currents in a conductor which are so high above the conductor rating that if allowed to continue, will permanently damage the conductor or equipment, must be interrupted before they become short circuits or high level ground faults. Fuses and circuit breakers provide the needed overcurrent protection.

"**Grounding**" is best divided into two elements: Service Grounding and Equipment Grounding and Bonding.

Service Grounding

This is the most confusing and misunderstood portion of the "grounding" system. All electrical systems do not come under the National Electrical Code, and the Code does not require that all systems that do come under the Code be grounded. However, the vast majority of the electrical systems we are dealing with in this text are "uni–grounded" systems, meaning that there is only one ground point at the service, and the system is installed in conformance with the Code.

Grounding one conductor of the system at the service insures that one conductor will be at ground potential and that the voltage relationship between the ungrounded phase conductors and the grounded conductor will remain unchanged, regardless of the effects on the system voltage by lightning or switching

surges. It should be stressed that grounding is not intended
establish a low impedance path for fault current through the
earth to the source generator or transformer.

Equipment Grounding and Bonding

First, bonding is the tying together electrically of metallic su
faces to limit the possibility that the surfaces might be ene
gized at different voltages, causing a shock hazard for anyon
who might touch the two surfaces simultaneously. This is a ve
important safety consideration.

Equipment Grounding is extending the bonding connectio
between metallic surfaces directly to the system grounded co
ductor and/or the grounding electrode conductor by means
an equipment grounding conductor. In addition to limiting an
difference in voltage between exposed metal parts and groun
potential, this conductor, which might be a separate conduct
or a metallic raceway, has another very important functio
When there is a fault between an energized conductor an
metal, the equipment grounding conductor assures that the
is a low impedance path back to the origin of the circuitr
namely the point where the neutral or grounded circuit condu
tor is connected to the grounding electrode. This completes t
low impedance path, causing a ground fault to become a hig
current level ground fault. This in turn causes the overcurre
device, either a fuse or circuit breaker, to interrupt the fault cu
rent in the shortest possible time.

If the impedance of the ground fault path is so high that t
fault current is below the activation level of the fuse or circu
breaker, the fault will persist until power is cut off or the fau
burns clear. This is the need addressed by equipment grou
fault protection. More on this later. It is clear that the effectiv
ness of grounding is only as good as every link in the chain, fro
the point of the fault to the point where the equipment groun
ing conductor ties to the neutral, or grounded circuit condu
tor. It also depends on the operation of the overcurrent device
interrupt the fault.

Any broken link, and the whole grounding safety advantage is lost. Even when the grounding circuitry is completely tested for integrity on a new installation, and it often isn't tested, there are many locations where time, usage, and abuse can cause a broken link. Also, the gap in the grounding system is unlikely to be detected because the grounding conductor only carries current when it is needed during a fault condition.

Some of the more likely locations where grounding continuity can be lost are:

• Grounding contacts of grounding receptacles.

• Use of 3-wire to 2-wire adapters. The pig-tail type is a special hazard because the pig-tail is frequently not attached to the wallplate screw and can touch the energized plug blade, energizing all the metal on the load side of the adapter. The adapters with a tab rather than a pig-tail give the false sense of grounding security, because the screw to which the tab is intended to be connected is usually connected to a device box which is not connected to a grounding circuit. This is because the receptacle is of the old 2-wire type, or else an adapter would not have been used in the first place.

• Two-wire cords. If a 2-wire extension cord is plugged into a grounding receptacle, and then an appliance with a 3-wire grounding plug is plugged into the 2-wire cord, with or without an adapter, grounding is defeated.

• Grounding pin removal. Frequently, the grounding pin is broken off a plug so it can be used directly in a 2-wire receptacle or cordset.

• Metal raceways and fittings. When the metal

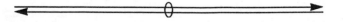

raceway is used as the grounding conductor as allowed by the National Electrical Code, every conduit or tubing joint, every box connection, and every fitting connection in the grounding path is subject to losing its conductivity through corrosion, abuse, high fault currents, etc.

There are fatalities that would not have occurred if an equipment grounding conductor had not been present. The author has served as an expert witness in several litigations where there have been such fatalities. Typically there are gross violations of the NEC and of common sense safety precautions, resulting in a gap in the equipment grounding conductor. The grounding conductor then becomes energized on the load side of the gap, automatically energizing the exposed metal, usually the housing of a tool. The person holding the tool is killed.

One case in particular had most of the "horror stories" rolled into one. An outdoor grounding–type receptacle was supplied through a GFCI receptacle. A homemade cord set consisting of flat 3–conductor SPT–3 cord, with a molded plug on one end and a duplex receptacle in a "jiffy box" on the other end, was frequently plugged into the outdoor receptacle. This flat cord construction is suitable for, and listed only for, indoor use. The grounding pin had been broken off the plug.

The cord had numerous abrasions, splices, and slices. One splice was partially covered with tape. Bare wire strands of all three conductors could be seen at more than one location. The "jiffy box," a small steel device box designed only for permanent surface mounting indoors, had no cord grip where the cord entered and had numerous holes for nails and conduits.

The GFCI receptacle, doing its job, had tripped repeatedly and was removed.

On the day of the accident, the cord and "jiffy box" were lying in mud and water. The victim plugged in an appliance with a very good equipment grounding conductor and was killed. It is likely that the equipment grounding conductor in the cord set had become energized, energizing the case of the tool, and allowing electricity to travel through the person's body to the wet ground.

Grounding: Where installed and maintained properly, it is very good. However, where it isn't, it can have horrible consequences. "When it's good, it's very, very good–and when it's bad, it's horrid!"

To summarize "grounding", there are key points to remember:

POINT 1 – Service grounding refers to grounding one of the electrical system's normal current–carrying conductors to the grounding electrode. The conductor grounded in this way is referred to as a " grounded circuit" conductor and may be a "neutral". It is usually identified by the color white. Service grounding keeps the system voltages stable relative to the earth and to each other, even during lightning and supply voltage surges.

POINT 2 – Equipment grounding and bonding refers to tying exposed metal surfaces together electrically and connecting them through an effective electrical conductor to the grounding electrode. The "effective electrical conductor," when an insulated or covered conductor, is identified by the color green, or green with one or more yellow stripes. The conductor can also be a bare conductor, a metallic raceway system such as rigid metal conduit (RMC), intermediate metal conduit (IMC), electrical metallic tubing (EMT), or other means permitted by the National Electrical Code. This is called the "equipment grounding conductor." This grounding conductor is not a part of the normal current–carrying circuitry, and carries current only during a fault condition.

Bonding minimizes differences in voltage between exposed metal surfaces, and the equipment grounding conductor makes it more likely that any ground fault will become a high level ground fault, activating an overcurrent protective device.

Chapter 4

European Grounding Practices and Systems

Until the post-World War II era, in the USA, protection for people and property from the harmful effects of electricity consisted of isolation, insulation, overcurrent protection, and grounding. The principles of residual current protection were known in the technical community, but the applications were confined essentially to electric utility generation, transmission, and distribution supply facilities.

Residual current devices were first applied to domestic uses in South Africa and Europe. There are many different electrical systems used throughout the world, and the systems tend to follow the colonial empires under which the particular countries had their origin. The countries which were a part of the British Empire tend to have electrical systems like those in England. Those which came out of German or French colonial status follow the electrical practices in those countries, etc. A survey of the systems used in 148 countries is included in the Appendix, courtesy of the Canadian Standards Association.

Comparing residential systems in Europe with those in North America, the first difference is the frequency– 50 Hertz (cycles per second) in Europe, and 60 Hertz (Hz) in North America. This difference is minor for the subjects we will be discussing.

A major difference is the utilization voltage in homes. Most homes in Europe are served with a 3-phase, 4-wire supply, at voltages ranging from 380 Y/220 v, 400Y/230 v, to 416Y/240 v. This means that at a typical wall receptacle outlet, which they call a "socket outlet", the voltage to ground (called voltage to earth by Europeans) is 220, 230, or 240v.

Domestic system voltages in North America are essentially half those in Europe. Most homes receive power through a 1-phase, 3-wire supply at 240X120 v. Some high rise dwelling buildings and those served from downtown networks have 208Y/120 v services. Voltage to ground is 120 v.

The big difference is that the double voltage in European systems means that half the ampere capacity is required for the same amount of power. The savings in conductor and raceway costs are tremendous. However, other things being equal, the risks of electrocutions and fires are greater at higher voltage. The truth is that "other things are not equal," and the end result is that the European systems exhibit essentially the same safety record as the North American systems.

One reason for this is the relative self–discipline of the people. The 220 v at the interface with the user is much less forgiving than 120 v. The Europeans respect it. As an example, Europeans know better than to drive a nail into a wall over a socket–outlet to hang a picture, because they know that an electrical conductor is most likely buried in the wall just below the surface running vertically to feed the outlet. Can you imagine trying to impose such self–discipline on people in the USA?

The International Electrotechnical Commission (IEC) headquartered in Geneva, Switzerland, has been the primary organization responsible for creating international electrical standards. The USA has participated in the IEC for many decades, with varying degrees of involvement.

In the IEC committee activity for electrical building system products, the USA has been a peripheral player until the last several decades. The leaders have been the Germans, together with the French and the British. The South Africans have made significant contributions in specific areas, including the residual current device standards activity. The result has been as one would expect. Standards that are intended to cover world–wide practices are heavily biased towards European practices, which are heavily biased towards German practices.

At one time when the author was with a large US manufacturer and was actively involved in IEC committee work, we calculated that Germany had 10 times the manpower the US had in IEC activity. These people were very skilled and technically competent, multilingual engineers. It should not be surprising then that the IEC standards are very similar to the German electrical standards.

A brief examination of the most common residential systems used in Germany and throughout Continental Europe, is helpful in understanding the similarities and differences between the North American and European systems. The IEC terminology is used in describing the systems :

TN System – This is a system "having one or more points of its source earthed, the exposed conductive parts of the installation being connected to that point by protective conductors."

The North American version of the TN system is a uni-grounded system, meaning that there is only one point where the neutral or grounded circuit conductor is grounded. In the European type of "wye" TN system, it is the neutral that is grounded and this is accomplished at the utility pole. This is the biggest difference with the North American system, which is basically a TN system, and which must be grounded at the service, independent of any utility grounding. European system bonding is similar to US practices, and there is a conductor to carry the ground fault current.

There are three variations of the TN system, the TN-C (Fig. 9A), the TN-S (Fig. 9B), and the TN-C -S (Fig. 9C):

> TN-C: This system has the grounded–circuit or system neutral "combined" with the equipment grounding protective conductor in a single conductor throughout the system. It is thus a "multi- grounded neutral" system. There is no USA National Electrical Code usage of this system. It is common with US utilities.

> TN-S: This system has a "separate" grounded–circuit or system neutral conductor from the equipment grounding protective conductor throughout the system. It is a "uni–grounded neutral" system and is

Figure 9A

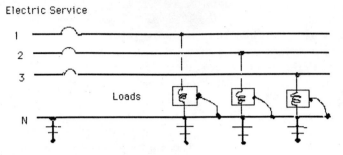

TN-C System
A Multi- Grounded Neutral System
No USA N E Code Usage

Figure 9B

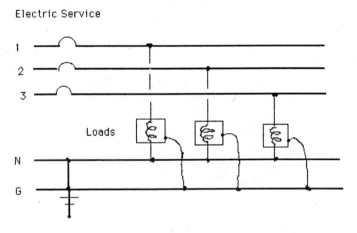

TN-S System
3 - Phase System Shown
Also Used As 1 - Phase, 2- Wire and 3- Wire
Systems with Ground

Figure 9C

Electric Service

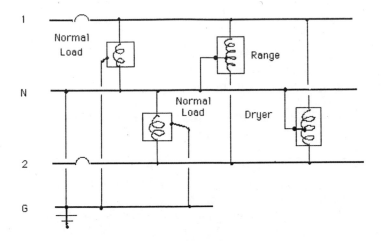

TN - C - S System

the standard system in North America for commercial and industrial installations, plus mobile homes and recreational vehicles.

TN–C–S: This system has a "common" grounded circuit or system neutral conductor and equipment grounding conductor for a portion of the system, and has a "separate" grounded circuit or system neutral conductor and equipment grounding conductor for the rest of the system.

This is our most commonly used dwelling wiring system, where the electric range and dryer circuits are TN-C circuits, and the rest of the building is TN-S. The range and dryer branch circuits have a

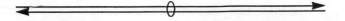

grounded–circuit conductor, which also serves as the equipment grounding conductor, as permitted in Section 250–60 of our National Electrical Code, where it was permitted for new construction up until the 1996 NEC. The 1996 NEC restricted it to existing installations.

It has been said that this special exception to the sacred rules of grounding was instituted to accommodate attempts by the electric utilities to make electric ranges and dryers cost– competitive with gas. The need during World War II to minimize the use of copper was also a factor.

TT System – This system has the ground at the utility pole as in the TN System, but relies on the path through the earth to carry fault currents. A separate ground is established wherever utilization equipment exists, and there is no conductor, other than the Earth, to carry fault currents. The TT system is illustrated in Fig. 10.

There are other IEC systems, but a discussion of them is not germane to this text.

Figure 10

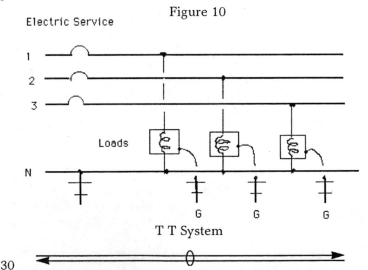

T T System

Chapter 5

TT System Leads To Need For RCD's - USA TV Problems

The big difference in electrical systems between Europe and North America is the introduction and widespread use of the so-called TT System. From conversations with my colleagues in the IEC, it would appear that the system was first introduced in Germany in response to pressure from the electric utilities, to follow electric utility practices and to minimize the cost of wiring.

For building use, the TT System is inherently unsafe for reasons I will explain, and would be totally unacceptable in North America. The Germans, however, instead of abandoning the system in favor of an inherently safe system such as one of the TN systems, used their ingenuity to invent products which made the TT system tolerable. It is this effort which triggered the revolution which I will call the residual current device revolution.

First, we must look at the inherent weaknesses in the TT System.

The TT System requires that the system be grounded at the service, in their case at the utility pole, that it be grounded to earth at utilization points, and that it rely on a low impedance through the earth to provide the only path for ground fault currents to activate overcurrent devices.

Well, we know that it just won't happen all the time. Even if one measures a low *resistance* value at the time of installation between the two grounding points, there can still be a problem. During a ground fault condition, there is the matter of *inductive reactance*, leading to such a high *impedance* that frequently there will never be enough current to activate a normal overcurrent device. Unfortunately, the ground fault current is high enough to do serious damage to people, equipment, and buildings.

An early attempt by the Europeans to solve this problem was

to invent the "4 X breaker." This is a circuit breaker that trip
instantaneously at 4 times its rating. Thus a 16 A breaker (a sta
dard European rating) would trip instantaneously at 64 A.
contrast to this, typical 15 and 20 A US breakers might requi
150 to 300 A to trip instantaneously. Of course, the low
trip values of the 4 X breakers causes them be more like
to experience "nuisance trips" during motor starting. Probab
this is a good time to mention some of the other basic d
ferences between European and North American circuit brea
ers, because the system parameters which affect circuit brea
ers relate to many other products we shall be discussin

We mentioned that the European conductors could be muc
smaller for the same kVA load because of the higher voltag
The result of the smaller European wires was higher impe
ance, leading to lower available fault current values whic
were required to be interrupted by the breakers. Whe
USA breakers had to have interrupting ratings of 10 kA ar
22 kA, European breakers could have much lower value

Another big point was that the electric services to th
homes in Europe have overcurrent protection of the wir
entering the homes, and we don't in the U.S.A. Current-lir
iting fuses, sized to minimize undesirable overload curre
interruptions, protect the service conductors and limit th
interrupting current the breakers will have to contend wit

The result was high precision "Swiss watch" 4 X breaker
which also have current limiting capabilities. The breake
were great, but the basic TT problems remained. The ne
development introduced the age of residual current protectio

The Europeans applied the principle that if you run a
the conductors which normally carry current through a don
shaped current transformer, regardless of the load balance of th
conductors, the vectoral sum of the currents will always be ze
under normal operating conditions. When the currents do n
vectorially add up to zero, then a current is going where yc
don't want it to go. If that current can be detected and use
to trip the power off, a dangerous condition can be avoide

The Europeans first introduced devices which they called residual current circuit breakers, the earliest of which were electro-mechanical devices without overcurrent protection, for installation at the dwelling electric service panel. The first units, after an abortive attempt at voltage sensitive units, were designed to trip at 500 milliamperes, or 0.5 A of residual current. The mechanisms were high precision mechanical units with highly sensitive micro-gaps. Later units tripped at 100 mA, and most units now trip at 30 mA.

Because of their sensitivity, these devices allayed the problems of interrupting ground faults on TT Systems. Other variations were introduced including units with overcurrent protection, receptacle or "socket-outlet" types, plug types, cord set types, and portable types.

Another inherent weakness with the TT System continues to cause problems. During voltage surges that can be caused by switching surges or problems on the utility lines, or lightning, the portions of the building system that are tied to the service ground will fluctuate up and down in voltage with the surges. As long as all portions of the system go up and down in voltage together, the problems are minimal, because the voltage differences are minimal.

However, where some portions are connected only to other grounds, large differences in voltages on exposed metal surfaces can result. Shock hazards, electric fires, and damaged electrical equipment can result.

A good example of how this problem can occur is the bad experiences in the USA when cable TV was first introduced. Many cables were installed to homes by cable TV people who were not licensed electricians, and did not follow National Electrical Code grounding requirements. Electrical inspections usually were not conducted.

A cable TV installer would run the cable to a point on the house, often remote from where the electrical service entered, connect the TV to the cable, and would drive a ground rod outside and connect the cable sheath to the ground rod. Inside the house, the receptacle supplying the TV had an equipment grounding conductor and a grounded circuit conductor, the white wire, both connected to the grounding electrode at the electric service, some distance away. If the TV had exposed metal surfaces and a three-wire plug and power supply cord, large differences in voltage could occur between the cable sheath and the exposed metal, particularly when voltage surges occurred. If the TV had a 2-wire plug with polarized blade, similar voltage differentials could occur between the cable sheath and any circuit component connected to the white wire.

We can use a hypothetical case to illustrate the problems. Please refer to Fig. 11, which illustrates grounding installed in accordance with the TT System.

Let's assume that when the electric service was installed, the electrician measured 25 ohms to ground in the grounding electrode circuit. For USA installations, Section 250-56 of the 1999 NEC sanctions this as adequate.

If a lightning surge on the power lines produced a surge current of 1000 A, not an unreasonable value, which came down the grounded circuit conductor and through the 25 ohms earth resistance, the voltage rise would be: $V = IR = 1000 \times 25 = 25,000$ volts.

If the TV cable sheath had only a separate ground connection, this voltage could appear across any two separately grounded surfaces at the TV set. We actually had an epidemic of TV failures and fires in the USA because of this unauthorized and illegal application of the TT System.

The solution is that the cable TV sheath must be directly grounded through an acceptable equipment grounding conductor, to the service grounding electrode. There should be no grounding rod at the cable TV entrance, unless it is electrically connected to the service grounding electrode. This makes the installation a legitimate TN-S installation, and solves the problem.

Figure 11

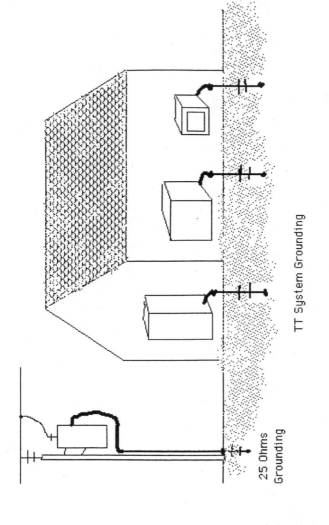

TT System Grounding

25 Ohms
Grounding

Example : With 25 Ohms Grounding Resistance, and a 1000 A
Lightning Surge, the Voltage Rise is :

E = IR = 1000 X 25 = 25,000 Volts.

Chapter 6
Early Medical Research

Prof. Theodore Bernstein of the University of Wisconsin has investigated the early medical research related to the effects of electricity on people, and has published several papers on the subject. He has graciously granted us permission to use information from his work. Two papers we are referencing are "A Grand Success," from the Feb. 1973 IEEE Spectrum, and "Theories of the Causes of Death from Electricity in the Late Nineteenth Century," in the Dec. 1975 issue of Medical Instrumentation.

In 1882, Edison completed the first commercial power station, the Pearl St. Station, in New York City. His DC system supplying incandescent lighting was an instant success. Before then, there was limited use of both DC and AC to supply arc lamps for street lighting. The AC system had the advantage that it equalized the erosion of the two electrodes. For AC to progress, it was necessary to invent an AC motor and a transformer. In 1886, William Stanley developed a practical transformer, and in 1888, Nicola Tesla invented the AC induction motor. Westinghouse obtained the patent rights to both, and a practical AC system was off and running.

The pressure on Edison increased, from the fact that about that time in 1888, a French syndicate cornered a large portion of the world copper supply and pushed up the price. The much higher copper content of the DC system versus the AC system became most important. Edison struck out to convince people that AC was unsafe.

With the accelerating growth of both DC and AC electric lighting systems from 1880 to 1900, there was a resultant increase in fatalities from electric shock. The medical profession became involved in many studies to understand electrical injuries, and how to prevent them.

The rivalry between the Edison DC system and the Westinghouse AC system, fueled by Edison's claims that AC was inherently "unsafe," increased the demand for information on the effects of electricity on people. Early studies using animals attempted to evaluate the effects of DC versus AC and the effects of various current intensities. Human cadavers were studied when available. Edison even attempted to have States pass laws limiting voltages to 800 v for DC, 550 v for pulsating DC, and 200 v for AC. Fortunately, he was not successful.

We mentioned in Chapter 1 the unfortunate role which William Kemmler played in the battle between Edison with his DC system and Westinghouse with his AC system. Kemmler was sentenced to become the first person to be executed in an electric chair, scheduled for June, 1889.

His defense attorney, former Congressman Bourke Cockran, appealed the sentence based on cruel and unusual punishment. This led to more experiments, investigations, and expert testimony, but the sentence stood, and Kemmler was electrocuted on Aug. 6, 1890.

Unfortunately, little scientific information was gained from his electrocution. Of the 25 witnesses, 14 were physicians hoping to gain knowledge. There was a separate meter and switch room from the electric chair, and no readings of current or voltage were taken. The 17 second first jolt, followed by a 70 second jolt, with smoke observed, would indicate a most inefficient procedure, to put it mildly.

Legal electrocution spread to 20 states, with all kinds of horror stories. The medical knowledge gained was minimal. The current values of 2 to 7 A applied for 2 minutes, and the current paths used were unrelated to most "real world" conditions. We know now that this procedure could cause cardiac asystole or "standstill", rather than ventricular fibrillation. When the current is turned off, the heart could possibly resume beating spontaneously.

Several plusses did come out of all this effort. The testing showed that 1000 ohms was a reasonable number for the human body impedance, under certain defined conditions, such as hand–to–hand. This value is in the ball park. More on this in

Chapter 7.

Before 1899, there were many strange theories as to how electricity caused death, including the following :

"Electric current deprives the arterial blood of its active magnetic properties and prevents the corpuscles from constantly inducing magnetoelectric currents in nerves."

"Normal atomic electrical equilibrium was disturbed by electric current to such an extent as to cause complete and instant death."

"In the legal electrocution of criminals, all the blood is forced to the head, resulting in death."

"Electric current constricts the arteries so that the heart cannot pump against the restriction."

In 1899 two independent studies were completed and the results were published at about the same time, by Prevost and Batelli at the University of Geneva, Switzerland, and Cunningham at Columbia University. These studies showed that the usual cause of death from low voltage shocks was ventricular fibrillation. We now know that since the effects of electricity are related to current, and not voltage, the terms "high and low voltage" shocks really refer to the system contacted, and the effects on people are really from "high and low current" shocks. Prevost and Batelli explained the possible use of electric defibrillators. It is unfortunate that such devices were not developed for use in medicine until 1947.

Dr. Kouwenhoven of Johns Hopkins Hospital is one of our greatest contributors to knowledge on electric shock. The importance of the Prevost and Batelli study is highlighted by the fact that Kouwenhoven has described how a translation of their work, given to him in 1930, started his group working to develop an electric defibrillator.

The Effects of Current on the Human Body

The earliest means of protecting people from electricity were the obvious ones: ISOLATION ("Stay away from it") and INSU-LATION ("Build a barrier between the bare conductors and people"). With the system introduced by Tom Edison came OVERCURRENT PROTECTION– the Fuse, and later the Circuit Breaker. GROUNDING was introduced in the pre–World War II era.

In order to make significant progress beyond these four effective means for protecting people from electricity, it was necessary to have a better understanding of just what the effects of electricity were on people. Electrical factors to consider are: differences in voltage, current levels, DC versus AC, various frequencies of AC, length of time of exposure, various wave forms of current, etc.

More information was needed on the resistance and impedance of the human body in its many sizes, shapes, and ages, the effects of various current paths through the body, the effects of moisture and other environmental factors, and detailed information about the body components, including the heart. Let's consider what makes the heart "tick."

First off, to paraphrase a commercial on TV, I'm not a medical doctor, and I don't even play one on TV. As a lay–person who has been exposed to many meetings among medical people and engineers on the subject, and the related literature, I believe I do have a layman's understanding. I was the USA Technical Advisor for many years on several International Electrotechnical Commission (IEC) committees which dealt with electrocution and how to prevent it. The most germane to this discussion was TC 64, Working Group 4, which had responsibility for the effects of current on the human body. On this working group were both engineers and medical doctors from around the world.

At the risk of oversimplifying, I'll take a shot at describing for us non-medical people how the heart works. Our body generates a tiny electrical current pulse, the "pacemaker current" that tells the heart when to beat. We've all heard of the artificial pacemaker, which supplements or substitutes for the real thing. The natural pacemaker current is a tiny, tiny current on the order of millionths of an ampere.

One of the best source documents is IEC Publication 479 – Effects of Current Passing Through the Human Body. The Table of Contents is informative by itself and appears as follows:

Part 1: General Aspects:
Chapter 1: Electrical Impedance of the Human Body
Chapter 2: Effects of Alternating Current in the
    Range of 15 Hz to 100 Hz.
Chapter 3: Effects of Direct Current

Part 2: Special Aspects:
Chapter 4: Effects of Alternating Current with
    Frequencies Above 100 Hz
Chapter 5: Effects of Special Waveforms of Current
Chapter 6: Effects of Unidirectional Single Impulse
    Currents of Short Duration

A typical death scenario from electric shock is as follows:

The heart beat cycle has a vulnerable period which is about 10% to 20% of the cardiac cycle. This corresponds to the first part of the "T- wave" in an electrocardiagram. Remember that the heart is receiving its normal signals from a tiny natural body current in the millionths of an ampere range.

When an electric shock current passes through the heart during the vulnerable period, it overwhelms the natural signals. Even though the shock current is likely to be relatively small, on the order of 100 to 300 milliamperes (0.100 to 0.300 A), it is about a thousand times larger than the normal body current.

The ventrical heart muscles get all confused as to what they are supposed to do, and just flutter–causing ventricular fibrillation. A heart will not come out of ventricular fibrillation by itself, and without help, the person dies in several minutes from lack of a oxygen in the blood supply.

We now know that the way to try to save the victim is to stop the heart completely, and encourage the natural pacemaker current to take over. It was not until 1959, that Dr. Kouwenhoven and his coworkers, building on the work of Prevost and Batelli, perfected a "closed–chest" defibrillator technique. Before then, it was usually necessary to open the chest cavity and manually massage the heart in an attempt to return it to its normal rhythm.

The defibrillator works best on a well–oxygenated fibrillating heart. Kouwenhoven was successful with closed–chest cardiac rhymic pressure and mouth–to–mouth resuscitation, coupled intermittently with the defibrillator.

It is clear that there are many variables in electrocution scenarios, and we shall touch on only a few in the following text. The above mentioned IEC document is of particular value for anyone interested in deeper knowledge.

The Person – Differences among people have a bearing on the effects of electricity. Most of the parameters set are based on "normal healthy male adults." Body size, weight, height, age, sex, and condition of health all have an effect.

Path of Current – Components of the body in the current path: skin, muscle, blood, bone, joints, tendons, tissue, organs, etc. represent an impedance consisting of a resistance and a capacitance. The values of the impedance and the resulting current depend on the current path, the touch voltage, the duration of the current flow, the frequency and wave form, the degree of moisture on the skin, the surface area of contact, and the skin temperature.

Skin impedance is a network of resistances and capacitances. The skin impedance value depends on voltage, frequency, time duration of current, surface area and pressure of contact, degree of moisture on skin, and temperature. Skin impedance decreases when the current increases. For touch voltages up to 50 v, skin impedance varies widely. Above 50 v the skin impedance decreases considerably and becomes negligible when the skin breaks down.

The impedance of the skin decreases when the frequency increases. The impedance of the rest of the body is resistive. Blood has a very low resistance, similar to that of salt water. Muscle tissue has a very low resistance when the muscle fibres are in parallel with the current path, and a higher resistance when the muscle fibres are perpendicular to the current path. Bones, except for the marrow, have high resistance.

Current Path through the Body – The route the current takes through the body affects the total impedance of the path, which affects the amount of current that will flow as a function of the voltage. " I" still equals "E/Z."

The "hand-to-hand" current path is taken as the norm, and if we assign a value 1.00 to it, or 100% body impedance, here are some of the relative body impedances for other paths:

Hand-to-hand	1.00
Hand-to-foot	1.00
Hand-to-head	0.50
Hand-to-chest	0.45
Hand-to-stomach	0.50
Hand-to-knee	0.70

In addition to the impedance of the current path, the relation of the path of the current to the heart also plays an important role in whether or not ventricular fibrillation occurs. Consideration is given to this by a term called the "heart current factor." If we take a path likely to cause ventricular fibrillation such as "left hand-to-feet" and assign a current likely to cause fibrillation, this factor will give us an idea of what currents through other paths are needed to present the same danger of fibrillation. Here are some heart current factors:

Path	Current
Left hand–to–either or both feet	100 milliamperes
Left hand–to–right hand	250
Right hand–to–either or both feet	125
Chest–to–left hand	67
Chest to–right hand	77
Back–to–left hand	143
Back–to–right hand	333
Seat–to–either or both Hands	143

Notice those chest–to–hand conditions. They're really dangerous. It is reported that the lowest accidental current that caused ventricular fibrillation was about 60 mA.

Frequency – The previous data are based on system frequencies of 50Hz and 60Hz, and are considered applicable for frequencies from 15 Hz to 100 Hz.

In general, if we were to select a frequency to be the most dangerous from a ventricular fibrillation standpoint, we couldn't have picked better than the 50–60 Hz range.

The lower you go, all the way to DC, the safer it is. For higher frequencies, the impedance of the skin varies inversely with the frequency. At 500 Hz, the skin impedance is only about 1/10th that at 50 Hz, and the total body impedance is essentially the internal impedance. In spite of this, the actual current necessary to induce ventricular fibrillation at 500 Hz is about 6 times what is required at 50 Hz. Actual data are sparse for the high frequencies.

An interesting point is the effect of lightning on people who are "struck" by lightning. Of course, if they are subjected to a direct lightning stroke, the current is so high that they suffer massive damage. But this is not the typical death scenario from lightning. The heart goes into ventricular fibrillation, similar to what happens in a typical building electrical system shock. This is caused by the much smaller current that goes through the person from the electrical field created near a lightning stroke.

To save the person, we must stop the heart from fibrillating.

Some people who are " struck by lightning" live and some die. The vulnerable period of the heart beat is approximately 1/7 th of the total heart beat cycle. The heart beat cycle time varies, but it can be considered as about 60 cycles per minute, or once per second for people at rest, and 150 cycles per minute or so for actively exercising people.

When I was active on the IEC and involved with the effects of electricity on the human body, we had information both from the USA and from Europe on the number of people who were killed by lightning versus the number who were " struck by lightning." Guess what! Both sets of data showed that one out of seven people who were "struck by lightning" actually died. Experience in Germany is that one person in three died.

We have concentrated so far on ventricular fibrillation, but other effects on people are worth covering such as atrial fibrillation, the effects of "perception current," "startle current," "let–go current," and effects on breathing.

Atrial Fibrillation – As I explained before, this text is not intended for a cardiologist, and I have greatly simplified what really goes on in the heart. We should mention that the heart has two arterial chambers and two ventricular chambers. The arterial chambers contract first, sending blood into the ventricular chambers. The left ventricular chamber pumps oxygen- rich blood to the body, and the other ventricle pumps blood needing oxygen to the lungs. Atrial fibrillation can occur from electric shock but is rarely fatal, without other complications.

Perception Current – This is the lowest value of current that a person can perceive. Most of the literature indicates that this value is 1 mA for 60 Hz current. However, one must look at the conditions under which the tests were done. The 1 mA was arrived at by having a person touch electrodes and turning on the current for 1/10th of a second. I believe this is unrealistic.

ANSI Standard C101 specifies 0.5 mA as the threshold and limit for startle reaction.

While I was with a large manufacturer of electrical equipment including GFCI's, I had a testing device which I called a "perception current tester" made into an attache case. Most of us have been shocked during our lives, but we never know what value of current we were receiving to relate it to the feeling. This tester allowed a person to place his left thumb and forefinger on two terminals, turn a knob with his right hand, and observe on a meter, the amount of current flowing through his hand while he was experiencing it.

For ten years I traveled the country extensively, and during that period I tested hundreds, if not thousands, of people. Women can sense lower currents than men, about 0.20 to 0.25 mA, and men can sense 0.30 mA. One milliampere is a heck of a jolt. The tester will go to 2.00 mA under normal conditions, but most people quit before 1.00 mA.

It should be mentioned that ALL the discussion in this text relates to electrical current actually conducted through a circuit or object, including people. The text does not address the subject of "EMF" – electrical - magnetic fields, and what effects they might have on people, if any. EMF has been receiving attention lately, both in the electrical industry and in medical circles, as well as in the press.

Startle Current – It must be emphasized that none of the GFCI's and similar devices we shall be discussing later limit the AMOUNT of current the person experiences. The basic laws of physics apply. The impedance of the person and the voltage across his body determine the current by our standard $E = IZ$ equation. It is the TIME to which the person is exposed that we can control, and we make the exposure time so short that ventricular fibrillation will most likely not occur.

This means that although we may save his life, he can get a powerful jolt. If we assume 1000 ohms body impedance with 120 v present, we get $I = E/Z = 120/1000 = 0.120$ A, or 120 mA. If you're on a ladder or in some other difficult orientation, this "startle current" could lead to injury.

In order to determine what level of leakage current or "startle current" can be tolerated, the ANSI C101 Committee engaged

UL to conduct a test program. The research committee for the test program, headed by Arnold Smoot of UL as Chairman, included such well known experts in the field of electric shock as Dr. Kouwenhoven of Johns Hopkins University and Professor Dalziel of the University of California at Berkeley.

Preliminary tests were conducted on 20 men and 20 women. The tests showed that women were more sensitive to low levels of current than men, and the tests proceeded with 20 women subjects. The tests led the C101 Committee to select 0.5 mA as the maximum allowable leakage current from exposed parts of cord–connected portable appliances.

Let–Go Current – This is the current value up to which a person can release an energized object, such as an electric drill. Test probes were used to gather "let–go" values for various segments of the population. All men and almost all women and children can let go up to 6 mA. At 10 mA 98.5% of men, 60% of women, and only 7.5% of children can let go. Only 7.5% of men, and no women or children can let go of 20 mA. No one can let go at the 30 mA level.

The results of the test programs were consistent with tests conducted years earlier by Baron Whitaker, retired President of UL, when he was a UL engineer. As a result of Whitaker's paper: "Shock as it Pertains to the Electrical Fence," UL acknowledged 5.0 mA as a safe level for leakage currents from appliances from the standpoint of " let–go" and fibrillation of the heart.

Because of many tests done on cross–sections of people measuring the current value up to which they could "let–go" of test probes, the minimum trip value for Class A GFCI's has been set at 6 mA. The GFCI must not trip below 4 mA, and must trip at 6 mA–a very tight window.

The European–type device most commonly used, referred to in IEC standards as an "RCD"–residual current device–has a minimum trip value of 30 mA. The Europeans do not consider "let–go" current as a safety consideration that needs to be prevented by RCD's. This is fortuitous, because the electro–mechanical type RCD really has difficulty operating reliably much below 30 mA.

The European devices with 30 mA trip levels are intended to protect approximately 90 percent of the human population. The North American devices with 6 mA trip levels are intended to protect approximately 99.5 percent of the human population - men, women, and children.

On the other hand, because of our low trip value, there are applications with "standing leakage" conditions above 6 mA where we cannot provide GFCI protection. Of course, the "standing leakage" should be as low as possible. As an example, the limits of allowable leakage current for cord- and plug-connected products listed by UL are 0.75 mA for equipment intended to be fastened in place or located in a dedicated space, such as household refrigerators and freezers. The limit is 0.5 mA for portable cord- and plug-connected equipment. UL Standard 943-Ground Fault Circuit Interrupters, requires that GFCI's must trip at 6 mA. Yet, three UL Standards allow equipment to have 3.5 mA of leakage current. These are:

UL 1244 – Electrical and Electronic Measuring and Testing Equipment.

UL 1950 – Safety of Information Technology Equipment, Including Electrical Business Equipment.

UL 3101-1 Electrical Equipment for Laboratory Use.

Part 1 – General Requirements.

The higher leakage current limits are allowed in order to satisfy FCC requirements by the use of electromagnetic interference filters, which can have high leakage current.

How can we expect a GFCI not to trip when two or more pieces of equipment covered by these 3 standards are supplied through the GFCI ?

I have many GFCI's in my home. My home office has various computers, printers, and other office equipment with FCC mandated filtering. The only circuit where I have had to remove the GFCI because of frequent tripping is the one that supplies my computer and printer. There is a need to harmonize allow-able electrical leakage current values.

Effects on Breathing – The current values between the "let–go" current and the ventricular fibrillation current–about 10 mA to 60 mA, can affect the ability to breathe. If a person were protected by a GFCI, the time would be so short before the current was turned off that breathing would not be affected.

Without GFCI protection, if a person were "frozen" to an AC electrical current for a long enough time without breathing, it could be fatal. It is likely that the current would increase, however, probably from body moisture, reducing the impedance, and possibly would cause the heart to go into ventricular fibrillation. Data on this are scanty.

To sum up, the best way to prevent people from being killed by electricity is to prevent their hearts from going into ventricular fibrillation. The heart will not come out of ventricular fibrillation spontaneously, and unless the person receives competent defribillation treatment promptly, death will result.

The physiological effects on the human body, and the way ground fault circuit interrupters protect people, are demonstrated in Figure 12. The GFCI will trip before a current can cause ventricular fibrillation. Note the other physiological effects in Zones 3 and 4 and how the GFCI trip curve protects people from those effects.

Figure 12

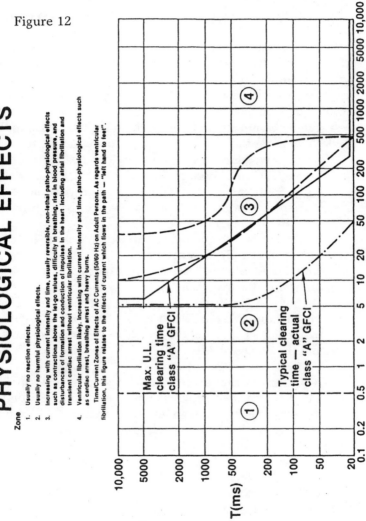

PHYSIOLOGICAL EFFECTS

Zone

1. Usually no reaction effects.

2. Usually no harmful physiological effects.

3. Increasing with current intensity and time, usually reversible, non-lethal patho-physiological effects such as contractions above the let-go values, difficulty in breathing, rise in blood pressure, and disturbances of formation and conduction of impulses in the heart including atrial fibrillation and transient cardiac arrest without ventricular fibrillation.

4. Ventricular fibrillation likely. Increasing with current intensity and time, patho-physiological effects such as cardiac arrest, breathing arrest and heavy burns.

Time/Current Zones of Effects of AC Currents (50/60 Hz) on Adult Persons. As regards ventricular fibrillation, this figure relates to the effects of current which flows in the path — "left hand to feet".

Chapter 8

International RCD Developments

I am indebted to many of my colleagues representing countries all over the world who participated with me on several IEC Working Groups, for the insight I have gained from them regarding the worldwide historical view of the evolution of residual current devices.

It would be impractical to name them all, but I wish to thank the following for their most helpful contributions to the text:

Mr. Viv Cohen, PE, currently Technical Services Manager, Circuit Breaker Industries, Ltd., Johannesburg, South Africa. I have paraphrased portions of his paper, "The Application and Usage of Earth Leakage Circuit Breakers."

Mr. Pat Ward, Managing Director, Western Automation Research & Development, Ltd., Galway, Ireland. I used information from his paper: "A Discourse on Residual Current Devices."

Prof. Dr. Gottfried Biegelmeier, Cooperative Testing Institute, Vienna, Austria. He provided me with many of his papers on the effects of electrical shocks and the RCD's of Austria.

Viv Cohen is mentioned first because it was in South Africa that the earliest progress was made with residual current devices. There was an attempt in the 1930's in the United Kingdom to perfect devices that would provide electric shock protection by detecting shock voltages, but the devices were discontinued after several years because of inherent limitations.

Electrical shock hazards were very severe in the gold mines of South Africa. Devices using cold cathode gas discharge tubes were introduced into the mines in 1956, operating at 525 v with a 250 mA trip value.

In the mid–1950's, West Germany introduced RCD's with trip values of 3 A down to 300 mA. Meantime, the transistor was born in the USA in the AT&T Bell Labs in 1948. The electronics revolution had begun. Back in South Africa, a device using magnetic amplifier technology was developed with a 20 mA trip value. Four hundred of these devices were installed in the homes of the mining village of Stilfontein in Western Transvaal in 1957 - 58. For more than a decade, South Africa led the World in the production and application of high sensitivity core–balance RCD's, and was the first country to require them on all 220 v receptacles.

How the Residual Current Device Works

This is a good time to explain as simply as possible just how these many devices that we are generically calling "residual current devices" or RCD's actually work. There are many names given to the devices which are included in the generic term RCD.

In the USA, we call the ones that are intended to protect people "ground fault circuit interrupters," or GFCI's. They are sometimes referred to as ground fault interrupters or GFI's. I must stress that these are the "people protectors." Two Classes of GFCI's are presently listed by UL. The Class A GFCI is the basic one, of which there are presently over 200 million installed in the USA. They have a minimum " must trip" value of 6 mA.

UL also lists a Class B GFCI with a 20 mA minimum trip value. Use of this device is rare and is restricted to underwater swimming– pool lighting fixtures installed before the 1965 National Electrical Code was published. Those lighting fixtures were allowed to have leakage current above 6 mA, and could cause Type A GFCI's to trip.

In the USA, we call the devices that give protection from system arcing ground faults, or low level ground faults, "ground fault protection of equipment," or "GFP" and sometimes "GFPE."

Internationally, the early terminology used was:

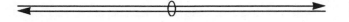

Earth leakage circuit breaker – ELCB
Residual current circuit breaker – RCCB
Residual current circuit breaker with overcurrent protection – RCCBO

The latest IEC International terminology uses the generic RCD, plus:

Socket– outlet residual current device – SRCD
Portable residual current device – PRCD
Residual current breaker with overcurrent pro tection – RCBO

Here's the principle behind all of them:

Fig. 13 shows a basic 2–wire circuit consisting of a black energized conductor and a white grounded–circuit conductor. The conductors have been run through a current transformer, or toroid, a kind of an electrical donut. Any current that runs

Figure 13

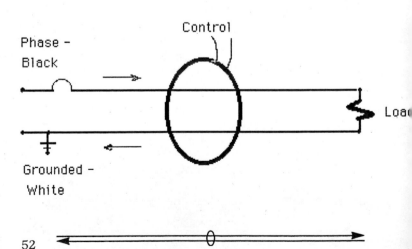

out through the black conductor, must return through the white conductor unless there is an electrical "leak" in the circuit. As long as there is no leakage of current, the currents are exactly equal and they have no effect on the current transformer.

It doesn't matter whether the circuit is a single phase, 2-wire circuit as shown, a 3-wire circuit, or a 3-phase, 4-wire circuit. It doesn't matter what the current balance of the loads among the conductors is. The NET current is always zero. This can be shown with vectors, but vector analysis is a separate subject and unnecessary for this discussion. Just remember that the summation of the currents is always zero unless some current is bypassing the current transformer. This current is likely a ground fault current.

I say "likely" to stress that all of these devices respond to RESIDUAL currents, and not just to ground fault currents. For example, they all have a test button on them, which is usually a button with a resistor connected to allow current to bypass the current transformer when the button is pressed, tripping the device. This creates a residual current condition.

This is why I have chosen to embrace the IEC term "RCD" for these devices, as a generic term rather than GFCI or GFP.

Let's return to what happens when a device detects a residual current above it's minimum trip level. This is where the major difference exists between the European and the USA types of RCD's. Although there are electronic types in usage in Europe, the European types are mostly electromechanical mechanisms. The USA types are electronic.

The current transformer is a transformer that has the conductors running through the donut as the primary, and coils forming the donut as the secondary. In Fig. 13, as long as the current going out through the black wire and the current returning through the white wire are exactly equal, the net current is zero, and no signal will be induced in the secondary coil.

When the returning current is less than the outgoing current by the minimum trip value, 30 mA for most European RCD's and 6 mA for USA GFCI's, the induced signal in the secondary causes the device to trip.

The European electromechanical RCD's are considered as "passive" or "voltage independent" type devices.

Pat Ward of Ireland describes the "passive" type best in his paper: "A Discourse on Residual Current Devices." To paraphrase Pat:

" The electromechanical type has a permanent magnet used to produce a holding force against a tripping spring. Under a non-fault condition, the magnetic force is stronger than the tripping force. Under a fault condition above the minimum trip value, sufficient magnetic flux opposing that in the permanent magnet will reduce the holding force of the permanent magnet so that the spring will take over, tripping the device. It is seen that this type will operate whether or not there is power supplied to the control circuit. Its operation is independent of line voltage."

The electromechanical types are high precision mechanisms with micro-gaps. They are particularly subject to corrosion and system voltage surges caused by lightning or switching. They are most vulnerable when installed on TT systems, as described in Chapter 5.

Chapter 9

The Birth of the GFCI

The person most credited with being the "father" of the USA–type of GFCI is Prof. Charles F. Dalziel of the University of California at Berkeley.

He had a wide and varied career, including stints at San Diego G & E and General Electric Co. He became well known for his work on electrical safety matters. He became involved in international electrical safety, and in 1961 he participated in a meeting of experts on electrical accidents in Geneva, Switzerland. He became exposed to the RCD work going on in Europe, South Africa, and elsewhere, and from his own testing was convinced that a lower trip level for RCDs was needed.

Upon his return to the USA he paired up with a west coast electronics company, Rucker Mfg. Co. An electronics technician, Reg Lester, who worked both at U. C. Berkeley and at Rucker, was instrumental in bringing Dalziel and Rucker together. A Rucker engineering team consisting of Bill Nestor, project manager, Elwood Douglas, electronics engineer, and Dick Doyle, electromechanical engineer, with Prof. Dalziel, developed the first electronic GFCI in 1961, and installed it in Prof. Dalziel's home in 1962. It was a 3–pole circuit breaker rated at 100 A, modified with a current transformer and electronics to trip at a residual current of 18 mA.

It is interesting that the first installation was at the service to the home and covered the entire home. Although this unit was successful, units with 6 mA trip levels would most likely not have been successful because of cumulative current leakage in the home and the fact that most homes use the grounded circuit conductor on ranges and clothes dryers in lieu of a separate equipment grounding conductor. This would cause the present generation of GFCIs to trip from a "double–grounded neutral" condition. We'll explain this in Chapter 10.

The emphasis shifted to incorporating GFCI capability into molded-case circuit breakers and into wall receptacles, and to lowering the nominal trip level to 5 mA, i.e. must not trip below 4 mA and must trip at 6 mA.

The first requirement for GFCI protection in the National Electrical Code was in the 1968 edition, in Section 680-4, to protect underwater lighting of swimming pools. The first GFCIs were "black box" self-contained units.

The first circuit breaker type GFCIs were introduced around 1968. Rucker Mfg. Co. worked with Pass and Seymour Mfg. Co. to develop the first receptacle type GFCI. P & S listed and introduced the first receptacle type GFCIs in Nov., 1972. Other manufacturers introduced them shortly after.

With increasing NEC requirements for GFCI protection, other companies were attracted to the manufacture of GFCI electronic modules and GFCIs. One of these, Electromagnetic Industries, made electronic modules for Square D and other companies and was later acquired by Square D.

Later, the business was purchased back from Square D by the original owners, to become Technology Research Corp.

For many years, and even now to some extent, the GFCI has been considered as only a backup to "good grounding." I cannot stress enough that GFCI protection is far superior to grounding as a means of protecting people from electric shock.

In 1982 I wrote an article on this subject that was published in the Jan. 1983 issue of the IAEI News magazine.

The management of the company I worked for at the time, which reviewed all my articles before they were approved, questioned my statements on the subject. I agreed to withdraw them if they would show me where I was incorrect. The article was published, and I don't think I can express myself any better on the subject now than I did then, so I'll quote from the article :

> "It's time for someone to speak up and set the record straight! The GFCI is more than TWICE as effective as equipment grounding in preventing fatal electric shocks from ground faults. Let's look at the facts: "

"There are basically two ground fault shock conditions which can result in death: series contact and parallel contact. The GFCI is extremely effective in preventing death in both series and parallel contact conditions."

"An equipment grounding conductor is TOTALLY INEFFECTIVE in the series contact condition. In fact, it can contribute to death by creating a path for the current to flow."

"In series contact, the person is the ONLY path for the current to flow to ground."

"How can a person be the only current path to ground? There are many ways, but two recent electrocutions will illustrate:"

"1. A baby sitting on a floor heating vent stuck a hairpin into a receptacle slot."

"2. A man operating a metal–encased electric drill with 3–wire cord and plug used a 3–wire extension cord supplied from a 2–wire receptacle through a pigtail adapter. The pigtail was not connected to the wallplate screw and contacted the line blade of the adapter, energizing the drill case."

The article goes on to explain that in parallel contact the effectiveness of the grounding conductor in preventing death is dependent on several variables, most notably the impedance of the grounding path compared with the person and the speed of the overcurrent device. In the parallel circuit condition, the current will divide in inverse proportion to the impedance of the person and the grounding conductor.

It would be an extremely rare condition that a person would be touching an energized conductor at the very instant a high level ground fault occurred in parallel with him. If it was in fact a low impedance ground fault path through the grounding conductor, the high ground fault current would have caused the overcurrent device to open the circuit before the person touched it. If the grounding path was so poor that the impedance was so high that the fault current was not high enough to operate the overcurrent device, then the person would receive a dangerous proportion of the ground fault current.

The statement is sometimes made that "GFCIs are no substitute for grounding." Let me paraphrase it: "grounding is no substitute for GFCIs." Good grounding has its place, and it makes a significant contribution to safety, particularly in overcoming the many hazards created by bad grounding.

True personnel protection requires that the person be protected from both the series contact and the parallel contact scenarios. If we want to protect people from both series and parallel contact scenarios, grounding won't do it–only GFCI's and similar devices will.

You only have to look in your kitchen to see one of the most glaring examples of the relative merits of grounding vs GFCI or residual current device protection. Kitchens have been required to have grounding–type receptacles for decades. How many appliances do you have around the kitchen with 3–wire plugs and cords on them? Practically all of them have 2–wire plugs and cords. Some appliances have all–plastic housings or are double-insulated. But look at the pop–up toaster. It usually has a metal housing or has exposed metal parts–and it only has a 2–wire plug and cord!

Why? The conventional wisdom is that when an oversized English muffin, donut, toast, or whatever is inserted and becomes jammed, people will tend to use a knife or fork to remove it. If the knife or fork should touch the bare heating element of the toaster when it is energized, and simultaneously touch the metal edge of the toast slot, if the toaster were to be grounded properly with a 3–wire cord, the contact would cause a high level ground fault, tripping the branch circuit breaker or blowing a fuse.

With the toaster ungrounded, as long as the person is not contacting ground, the person will be just like that bird sitting on a bare transmission line, and will be OK.

Wait a minute–we're not requiring grounding of the toaster because the grounding might do what it is intended to do, and we're relying on luck to prevent the person from receiving a dangerous shock? All the person has to do is contact grounded trim around the counter, a faucet, the dishwasher, the refrigerator, or any other ground, and we could have a fatality, unless there is GFCI protection present.

The arguement has been made to me that grounding is not a good idea for these appliances because if someone should touch an energized part with a utensil, NOT touch the grounded metal surface of the toaster with the same utensil, and yet simultaneously touch the grounded surface of the toaster, he could receive a fatal shock, lacking GFCI protection. This is a highly unlikely scenario, but possible. The same type of scenario can be applied to any equipment properly grounded. It's just like the lineman doing "hot line" work from an insulated bucket with a grounding conductor touching him. It is a good example of how grounding can contribute to death in the series contact condition, without GFCI protection.

The toaster should be designed so that the likely insertion of a kitchen utensil like a knife or fork cannot contact an energized surface. In addition, an equipment grounding conductor should be required, just like any appliance with exposed metal "likely to become energized" when there is a breakdown in the normal insulation. Product listing should include a test to insure that insertion of a kitchen utensil cannot create a shock hazard. It is of note that pop–up toasters in Europe, operating at twice the voltage as in the USA, have an equipment grounding, "earthing" conductor.

In any case, the GFCI must be doing something right. When I wrote the article back in 1982, I mentioned hundreds of thousands of GFCI's in use, and today we're quoting numbers in the 200 million–unit range.

Chapter 10

The Evolution of Various GFCI Types

The tripping characteristics and basic operation of the various types of GFCIs are pretty much the same. Fig. 14 demonstrates how a Class A GFCI protects a normal healthy adult from being "frozen" to a power source and from the usual death cause from electricity, ventricular fibrillation.

It should be stressed that all residual current devices, including GFCIs, require a "residual" current condition to exist in order for them to work. None of them will protect a person who comes in contact with just a live (black) conductor and a grounded circuit conductor-(white), or between two energized conductors. The person is just another load like a light bulb, with a current through him, likely similar to that current through a 15 watt to 40 watt light bulb, so far as a GFCI is concerned. No overcurrent device, fuse or circuit breaker, will operate, either, because there will be no "overcurrent."

Referring to Fig. 14, at 6 seconds of shock duration the center line of the band for the time–current curve for the "threshold" of ventricular fibrillation is shown as about 60 mA, with a minimum of 50 mA and a maximum of 70 mA. This covers 95% of "normal healthy adults." A more conservative value for the minimum current which can cause ventricular fibrillation with 6 second exposure in "normal healthy adults" is 35 mA.

The maximum time–current curve for UL–listed Class A GFCI's is a curve plotted from the formula: T in seconds equals 20 divided by current I in mA, and raised to the power 1.43, or $T = (20/I)^{1.43}$. For a minimum trip value of 6 mA, the time is about 5.6 seconds. For 20 mA of current $(20/20)^{1.43} = 1$, so the UL standard allows the GFCI to take 1 second to trip. The actual performance of listed Class A GFCI's is also plotted, and shows that they actually trip in 0.1 second maximum time at 6 mA, and less

Figure 14

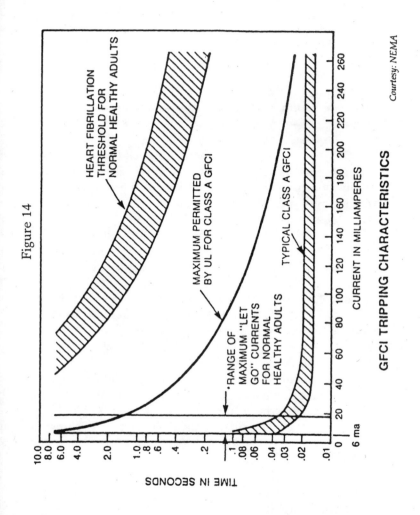

Courtesy: NEMA

GFCI TRIPPING CHARACTERISTICS

than 0.03 second at 20 mA.

Just think about this: when the current through a GFCI 10.000 A going to a load through the black wire, and only 9.99 A comes back through the white wire, the GFCI trips in 0.1 seconds! And all this miracle in a product available in a receptacle type at retail for considerably less than $10. It also has its own testing means, serves as a wall receptacle, and even comes with a wall plate. Of course, we expect it to last forever! I think this every time I shop for a birthday card or a Get Well card for $2.00 or so. A piece of paper-one look-"Isn't that nice!"-and throw it away!

There is an additional function that all listed GFCI's perform which most users are not familiar with. They constantly and continuously monitor the load side of the grounded circuit conductor to be sure that there is no grounding of this conductor on the load side of the GFCI.

GFCIs are to be used only on a grounded system, but the system is "uni-grounded," in which the single ground of the grounded–circuit conductor is at the service. The concern is that if a second ground occurs on the load side of the GFCI ground fault current could go through a person, return to the circuit through the second ground, and thence through the GFCI sensing current transformer. Thus the current could be going through a person, and the GFCI would not respond because there would be no residual current.

To eliminate this possibility, GFCIs have a second current transformer that constantly monitors the load side grounded circuit conductor (the white wire) and trips the GFCI if a ground occurs on the load side of the GFCI. Fig. 15 shows a simple GFCI circuit with the grounded neutral transformer, and Fig. 16 is a more detailed GFCI circuit.

This feature has made "honest" people out of some electricians who in the past would use the grounded circuit conductor, the white wire, to serve as the equipment grounding conductor also, in order to save a third wire. The GFCI trips under such conditions. What a sophisticated critter the GFCI is!

Figure 15

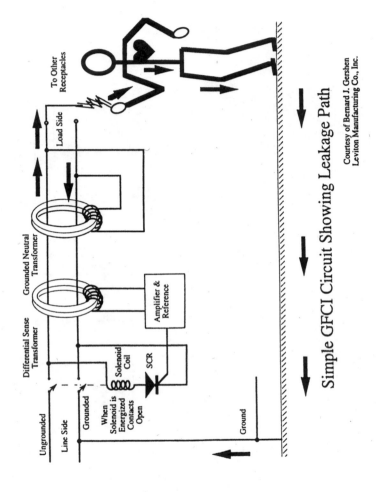

Simple GFCI Circuit Showing Leakage Path

Courtesy of Bernard J. Gershen
Leviton Manufacturing Co., Inc.

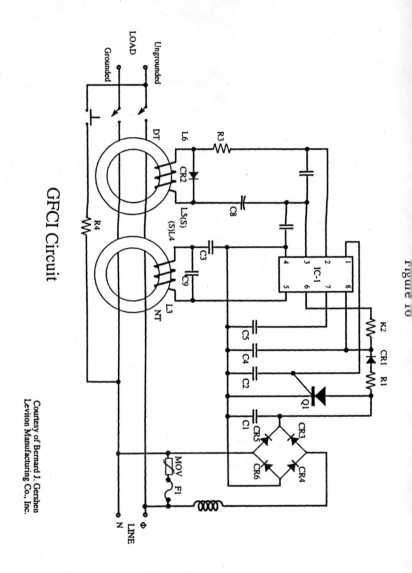

GFCI Circuit

Figure 10

Courtesy of Bernard J. Gershen
Leviton Manufacturing Co., Inc.

The present GFCI types are: self-contained, direct-wired or "black box" type, circuit breaker type, receptacle type, portable type, power supply cord or extension cord set type, and plug type.

Self-Contained Type

The first GFCIs were of this type, the direct offspring of the original one installed in Prof. Dalziel's home in 1962, direct-wired into the home's electric service, except with the trip level reduced to the 6 mA range.

One also appeared with an isolation transformer and a 0.5 mA trip level, but it was short-lived. The weight and cost of the transformer were major obstacles, and the trip level was too low.

Self-contained types are usually used for special applications such as high pressure water fountains, and applications where 3-phase units, or special voltage and current conditions exist. They are currently available in ratings up to 480 Y/277 v, 3 phase, 4-wire, 80 A. The 20 mA Class B GFCI still listed by UL for use on old swimming pool lighting is usually of this type.

The receptacle type unit, made with a blank face without outlets, is often used as a self-contained GFCI. Fig. 17 shows a self-contained type GFCI and Fig. 18 shows a blank face receptacle type GFCI.

Circuit Breaker Type

As soon as a clear requirement was defined for GFCIs, the circuit breaker manufacturing industry responded by developing a line of circuit breakers that also incorporates a Class A, 6 mA trip level GFCI into circuit breakers. The units can be substituted for standard circuit breakers in load centers or panelboards. They provide personnel shock protection, as well as overcurrent protection for the entire branch circuit fed from the unit.

Circuit breaker types are available from 15 A to 50 A, single pole, and up to 60 A in two pole units. Fig. 19 is a single pole

Figure 17
Self-Contained Type GFCI
Courtesy: Technology Research Corp.

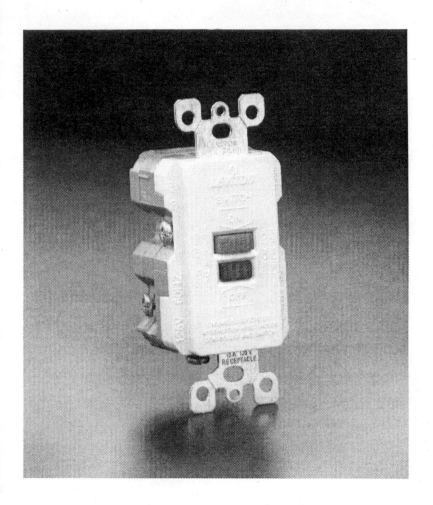

Figure 18
Blank Face Receptacle Type GFCI
Courtesy: Leviton Mfg. Co.

circuit breaker type GFCI, and Fig. 20 is a two pole unit. Fig. 2
is a schematic diagram of the single pole GFCI circuit breake
Note that the circuit breaker contacts interrupt only the ene
gized, normally black conductor. Normally the currents throug
the current transformer are equal, and no current is generated i
the secondary coil of the current transformer. As soon as thes
currents are unbalanced by 6 mA or more, the current tran
former sends a signal to the electronic "brain" of the GFCI. Th
current is magnified and is used to activate a solenoid, trippir
the breaker. Of course, only the GFCI capability of the unit
shown here.

The breaker can be tripped by overcurrents, as described i
Chapter 2. The unit, like all GFCIs, has an integral test circui
When the test button is pressed, a residual current of about
mA is introduced, which trips the unit. The breaker is reset i
the normal way of manually turning it to the "Off" position, ar
then to "On."

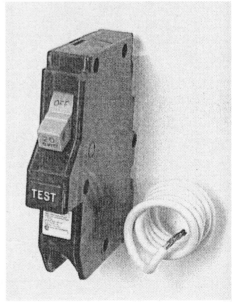

Figure 19
Single Pole
Circuit Breaker
Type GFCI

Figure 20
Two Pole Circuit Breaker
Type GFCI
Courtesy: Cutler - Hammer

One advantage of the circuit breaker type GFCI is that it provides protection for the entire branch circuit. Sometimes there is an excessive accumulation of current leakage on a branch circuit that could cause the GFCI to trip unnecessarily. Poor wiring, long circuit runs through wet environments, and old appliances, particularly those with heating elements in them, are sources of excessive current leakage. Of course, the faulty conditions should be corrected. Sometimes, the use of one or more receptacle type GFCIs on portions of the circuit will alleviate the tripping problems.

Receptacle Type

Fig. 22 is a schematic diagram of the receptacle type GFCI. In basic principle, it is similar to the other types. When it trips, it disconnects simultaneously both the energized black conductor and the white grounded circuit conductor. Also, in addition to the "test" button, it has a "reset" button. Note on the diagram

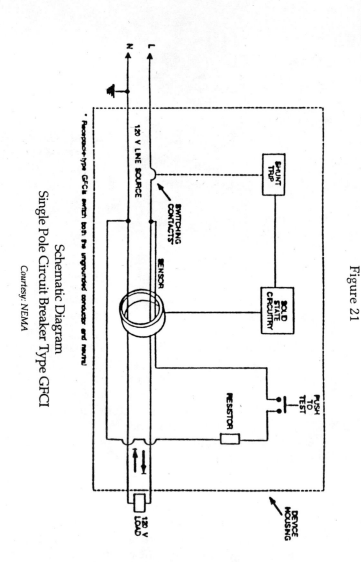

Figure 21

Schematic Diagram
Single Pole Circuit Breaker Type GFCI

Courtesy: NEMA

Figure 22

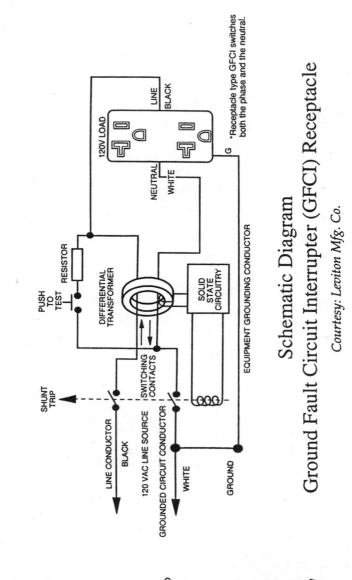

Schematic Diagram
Ground Fault Circuit Interrupter (GFCI) Receptacle

Courtesy: Leviton Mfg. Co.

that the equipment grounding conductor, usually bare or insulated green, plays no role in the operation of a GFCI. The equipment grounding conductor is connected to the "U" shaped grounding opening in the face of the receptacle. It runs to the electrical service, where it is connected to the building system grounding electrode and the system grounded circuit conductor.

It is clear that the GFCI operates independently of the equipment grounding conductor. This is more obvious in the case of the circuit breaker type, because there are no equipment grounding conductors connected to the circuit breaker. People become confused in the case of the receptacle type because the grounding conductor is present. It is only present to connect to the grounding contacts of the receptacle, just like any grounding type receptacle, and is independent of the GFCI function. Fig. 23 is a typical receptacle type GFCI.

All products experience changes over their useful lifetimes, often the result of field experiences. The receptacle type GFCI has had several improvements incorporated into the product and the packaging as a result of field experiences. It was observed that, in spite of the directions included with each device regarding the importance of proper wire connections, people were miswiring the GFCIs. The " Line" and " Load" connections are imprinted on the terminals, yet many devices were installed with the connections reversed. This results in GFCI protection still provided to the integral receptacles, but the feed- through, down stream loads are unprotected, because they become connected to the line side of the GFCI.

To minimize the likelihood of miswiring, the GFCI receptacles are now required to be shipped with tape over the Load terminals. The tape must be removed to make connections for feeding through to down – stream loads. The instructions have been improved also to stress the importance of correct connections.

Another common problem is the result of the human - nature tendency not to test the device monthly, as instructed. Some devices are NEVER tested. Nothing lasts forever, except "I and thou, and I'm not so sure about thou". All electrical system components are exposed to the possibility of overvoltages from lightning and switching surges. It is particularly important that electronic components not be subjected to voltages beyond their ratings. GFCIs should be tested monthly.

If a GFCI, when energized at 120v AC, does not trip when the test button is depressed, it should be replaced. However, since the integral receptacles and any receptacles fed from a GFCI receptacle are likely still to be energized, the tendency is not to replace the GFCI, and to continue to use the receptacles.

In response to this problem, Leviton Mfg. Co. has developed a GFCI receptacle that will automatically become deenergized if it is not providing protection. Pressing the reset button automatically tests the GFCI, and the unit will not reset if it is not functioning properly for any reason. If there is an open neutral on the line side, the GFCI will trip in response to the test button, but it cannot be reset.

To insure that the GFCI is installed properly, the device is shipped in the OFF position so that it must be reset upon first usage. If it is not correctly installed, it will not reset.

Portable Type

The portable type as the name indicates, can be easily moved about to provide GFCI protection to any device or cord set plugged into it. There are one or more receptacles integral with the portable GFCI to receive the plugs of the loads to be protected. The supply side of the portable type has either an integral plug or a power supply cord with a plug to connect to the electrical supply. GFCI protection is provided only from the integral receptacle to the load.

There is an additional requirement applied to listed portable type GFCIs that is not applied to some of the other types. Portable types are likely to be used on wiring systems of questionable integrity such as on temporary wiring and in the construction environment. Under these conditions, it is more likely that one of the circuit conductors could be broken on the supply side of the GFCI. If it is the

Figure 23
Receptacle Type GFCI Rated 20A, 125V With Indicating Light
Courtesy: Leviton Mfg. Co.

energized or black conductor that is broken, no hazard exists at the GFCI, and it is readily obvious that there is no power.

If it is the grounded circuit conductor, or white wire, that is broken on the line side of the GFCI, it would be obvious that an appliance would not work, but the line voltage terminals would be energized. Also, since the brain of the GFCI relies on a complete power supply in order to operate, a standard GFCI would not trip with a ground fault on its load side. A problem would be detected if a person tested the unit with the test button before use as per the instructions, but we know that some people will not do this.

Because of this concern, there is a requirement that the load terminals of portable GFCIs must be de-energized when the grounded circuit conductor is interrupted on the line side of the device. Most portable GFCIs accomplish this by having a 2-pole "normally open" relay in the line side. Thus, power must be complete to the relay in order for the contacts to be closed. If there is no power, such as from an open grounded circuit conductor, the relay contacts are opened by spring pressure. Power is necessary to overcome the spring pressure, closing the contacts. Fig. 24 is a typical Portable Type GFCI.

Power Supply Cord Type

The power supply cord type or extension cord type is similar to the portable type except that it is built into a cord. It is made to conform as closely as possible to the contour of the cord. The cord with such a GFCI looks like a python which has just eaten a pig.

On this type of GFCI, the test and reset buttons are usually protected by bosses to protect them from being depressed unintentionally when the cord is used. This type must meet the same requirements as the portable GFCI regarding open grounded circuit conductor operation. The load end of the cord must be de-energized when the grounded circuit conductor is interrupted on the line side of the device. The device is marked indicating that it should be tested before use, and testing would highlight any supply conductor problems. Fig. 25 shows a cord type GFCI.

Figure 24
Portable Type GFCI Rated 15A, 125V
Courtesy: Leviton Mfg. Co.

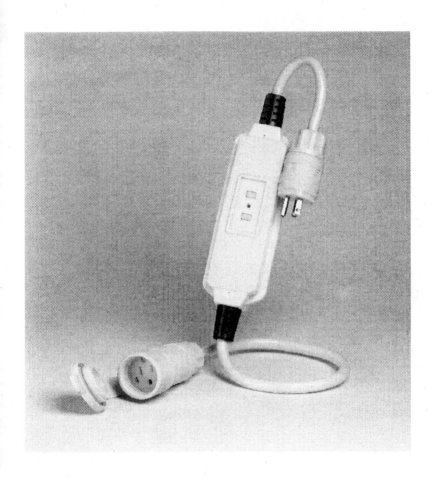

Figure 25
Cord Type GFCI
Courtesy: Leviton Mfg. Co.

Plug Type

Plug type GFCI's are available in several forms. Some are factory assembled to a power supply cord or extension cord set, some are made in the form of an adaptor, where they have both plug blades and one or more receptacle outlets, and some are available as a user–attachable unit. Plug type requirements are similar to those of portable and cord types. Fig. 26 is a plug type GFCI, and Fig. 27 is an adapter type GFCI.

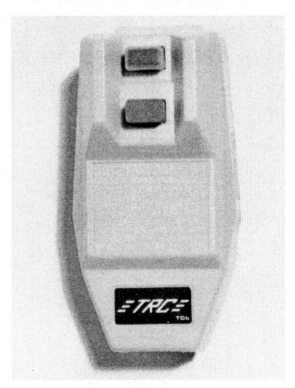

Figure 26

Plug Type GFCI
Courtesy: Technology Research Corp.

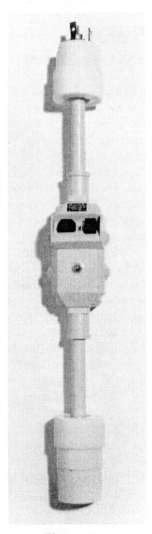

Figure 27
Adapter Type GFCI
Courtesy: Technology Research Corp.

Chapter 11

Evolutionary National Electrical Code Requirements for GFCI, GFP, and AFCI Protection

A bit of an explanation is in order for those readers who may not be familiar with the National Electrical Code and its continuous revision cycle.

The National Electrical Code, ANSI / NFPA 70, is an American National Standard sponsored since 1911 by the National Fire Protection Association. The original Code was created in 1897 as a result of the combined efforts of insurance, electrical, architectural, and related interests.

The acceptance and implementation of the National Electrical Code, referred to as the NE Code or NEC, and the requirements for products which had passed a testing process such as that provided by the "listing" service of Underwriters Laboratories (UL) was triggered by the response to the terrible Chicago Iroquois Theatre fire of 1903.

The theatre had electrically–illuminated exit lights at the exits, supplied through a disconnect switch. Because the exit lights were considered distracting during a low level lighting scene on the stage, they were turned off. After the scene ended, they were left off. The spotlights and floodlights were turned on. A very hot electric flood light on a platform illuminating the stage contacted drapery, and the drapery burst into flames. The theatre became dark except for light from the flames. Panic reigned. Nearly 600 people perished in the resulting conflagration. It should be noted that in the creation of the NEC, the original emphasis was on fire prevention, and the desire to minimize the loss of life and property damage caused by fire, and thus also to minimize insurance settlements.

The NEC is written so that it can be adopted by reference by State and local authorities responsible for laws, ordinances, regulations, administrative orders, or similar instruments. It is this adoption into mandatory requirements that is critical for the usage of products whose only function is safety. The GFCI, GFP, and similar devices probably would not have had such widespread usage and such success at preventing fatalities and property damage without mandatory requirements in the NEC.

GFCIs are now well known by the public and are sold directly through retail outlets in quantity. However, when GFCIs were first introduced, attempts to sell such a safety device directly were disappointing. Saul Rosenbaum, former VP of Research for Leviton Manufacturing Co., tells the story of his experience with their newly introduced receptacle–type GFCI. Saul set up a booth at a local Sears store and attempted to sell the GFCIs to customers at a special price. He pointed out the great safety features to many people. A typical response was something like: "Hey, it's a great product for the other guy, but I'm careful around electricity, and I don't need it." Saul sold two. Mandatory requirements are in order for items like seat belts and GFCIs.

For many years, the NEC has been on a 3–year code revision cycle. At the beginning of every cycle, proposals to revise the NEC are received by NFPA from anyone–from individuals as well as from organizations and corporations.

Responsibility for reviewing the proposals (usually about 4,000 per code cycle) leading to revision of the NEC, rests with 20 technical subcommittees called "code panels." Each code panel has responsibility for specific portions of the NEC. The code panels report to a "correlating committee" which collectively make up the National Electrical Code committee. The NFPA Standards Council oversees the actions of the National Electrical Code committee and appoints its members.

The code panels have balanced membership among all the interests involved including electrical inspectors, electric utilities, electrical manufacturers, testing laboratories, electrical contractors, unions, professional societies, trade associations, insurance interests, and user groups both large and small. Special experts are added as appropriate.

All proposals received by the NFPA by a given deadline are reviewed by the appropriate code panel at a meeting. A proposal must pass through a two-stage process in order to be included into the next edition of the Code.

At the first meeting in a code cycle each panel acts by a hand vote of the majority as to whether to: accept, accept in principle, accept in part, accept in principle in part, or reject each proposal under its jurisdiction. The panel may create its own proposals, and the action taken on panel-generated proposals must be to "accept."

After the meeting, a written ballot is sent to each panel member, and a signed ballot is returned. A two-thirds affirmative vote is necessary for any action that would change the NEC. All panel actions except "accept" must be accompanied with a panel statement explaining the reasons for not accepting the proposal as submitted.

The results of all 20 code panels are sent to the correlating committee for review. The correlating committee resolves any conflicts among the actions of the 20 panels and takes other appropriate action which may come under its responsibility.

All the proposals and a complete accounting of the actions taken are published in a document called the "Report on Proposals" or ROP. Previously this document was called the "Technical Committee Report" or TCR. Many years ago this was referred to as the "Preprint." Copies of the ROP are made available to the general public, and a copy is sent to every person who submitted a proposal.

A second stage in the code revision process follows with the receipt by a deadline date of "comments" on the proposals. Again, anyone may submit a comment on any or all proposals. The comments go through a similar process to the proposals. The 20 code panels meet to review and to vote on all comments.

The results go through the correlating committee, which must be doubly careful at this stage that whatever comes out results in a technically correct code without conflicting requirements. All the comments and actions taken are published in a document called the "Report on Comments" or–you guessed it–ROC. Copies of the ROC are made available to the general public, and a copy is sent to every person who submitted a comment.

A floor vote of the entire membership of the NFPA at the next annual meeting of NFPA confirms the revisions, and these are reviewed and subject to final acceptance by the NFPA Standards Council. What a busy 3–year cycle! The new NEC is printed and issued in the fall following the May annual meeting at which it is accepted. The local jurisdictions, cities, States, etc. must take action as desired to put the particular issue of the NEC into law, with or without local exceptions.

Now that the NEC code–making process is "perfectly clear" we can proceed to review the historical changes which have had such a significant effect on the growth of the use of residual current devices including GFCIs and GFP, and others such as the AFCI.

1968 NEC

The first requirement for GFCI protection in the National Electrical Code was in the 1968 edition. Article 680 of the code covering Swimming Pools, required in Section 680-4 (g) that underwater lighting fixtures present no shock hazard either by: 1) the design of the fixture, or 2) being protected by a GFCI. A definition of a GFCI was included.

It is to be noted that there is a considerable amount of work which has to be done to convert a code requirement for a new type of product into a third–party testing laboratory test standard in order for the test lab to evaluate the suitability of a product to satisfy the new code requirement.

The new GFCI swimming pool requirement is a good example. UL assumed the role of investigating appropriate GFCI test criteria. Arnold Smoot and Charles Bentel of UL exposed themselves to many actual electric shock tests in a tank of water to help set the product parameters. The tests resulted in establishment of a sound standard for GFCIs.

The results are covered in the IEEE paper 64-12: "Electric Shock Hazard of Underwater Swimming Pool Lighting Fixtures."

For those interested in tracing the documentation for this historic NEC change, it can be found on Page 213 of the 1968 "preprint," Proposal No. 43, from the Technical Subcommittee on Swimming Pools. Considerable testing with volunteers in an actual water environment supported the proposal.

1971 NEC

Section 210-7 required that all 15 A and 20 A receptacles on construction sites have GFCI protection, effective Jan. 1, 1974. The 1971 "preprint," on page 19, traces this requirement to Proposal No. 27 submitted by the National Electrical Manufacturers Association (NEMA).

Section 210-22(d) required that all 15 A and 20 A receptacles installed outdoors at dwelling– type occupancies have GFCI protection, effective Jan. 1, 1973. Page 21 of the 1971 "preprint" shows that this requirement originated with Proposals 39 and 40.

Section 215-8 specified that the GFCI protection could be on the feeders supplying 15 A and 20 A receptacles, in lieu of the provisions of Section 210-22(d). The 1971 "preprint" shows on pages 23 and 24 that this wording came from Proposals 51 and 52.

Section 230-95 required for the first time that ground fault protection of equipment (GFP) be installed for grounded wye electrical services of more than 150 v to ground, but not exceeding 600 v phase–to–phase for any service disconnecting means rated 1,000 A or more.

On pages 40 and 41 of the 1971 "preprint," Proposal 53 from NEMA explains the reasons for the new requirements. We discuss GFP in detail in Chapter 12.

Section 555-3 mentioned that GFCI protection of receptacles at marinas and boatyards would provide additional protection against line–to–ground shock hazards. On page 225, the 1971 "preprint" shows that this wording evolved from Proposal 52.

Section 680-6 required that no outdoor receptacles be located within 10 ft of the inside walls of a swimming pool, and that all receptacles located between 10 ft and 15 ft of the inside walls of the pool have GFCI protection. Page 244 of the 1971 "preprint" lists Proposal 87 as the source.

Section 680-31 required that all electrical equipment used with storable swimming pools be protected by GFCIs. The 1971 "preprint," on page 248, shows that this requirement evolved from Proposals 100 and 102.

1975 NEC

(Note: The NEC slipped a year from the 3–year cycle. What would have been the 1974 NEC became the 1975 NEC.)

Section 210-8(a) picked up the requirements for GFCI protection of outdoor receptacles at dwelling–type occupancies formerly in Section 210-22(d) and added requirements for GFCI protection of receptacles in bathrooms of residential occupancies. The 1974 " Preprint", Part II, on page 23, indicates that Proposals 36 and 39 initiated these requirements.

Also introduced into Section 210-8(a) was a definition of "bathroom." Believe it or not, this was a hot topic. We all thought we knew what a bathroom was until there were GFCI requirements associated with them. After that people attempted to avoid the GFCI requirements by declaring the area not to be a bathroom.

Section 210-8 (b) picked up the construction site GFCI requirements, with revisions, formerly in Section 210-7. The 1974 "preprint", Part II, on page 19 lists Proposal 15 as initiating this action.

Section 680-41(a) added requirements for GFCI protection of branch circuits supplying fountain equipment.

The 1974 "preprint", Part II, on page 268, shows that Proposal 37 led to this requirement. There have been some terrible accidents from electrical shocks around fountains. One is particularly engraved on my mind. Two couples, after a romantic dinner at a southern east coast resort, strolled to a decorative fountain. One girl removed her shoes and waded in the water. When she was in distress, her male companion jumped in to help her. The other two followed to help, and ALL FOUR were killed!

1978 NEC

Article 100 picked up the definition of "ground- fault circuit-interrupter" from Section 680-4(g), since the term was now being used in more than one Article. The 1978 "preprint" on page 8, Proposal 44, documents this change.

Section 210-8(a)(1) required that GFCI protection be provided at receptacles in dwelling unit garages, as well as in bathrooms. The 1978 "preprint" on page 32, Proposal 51 initiated the change.

Section 210-8(a)(2) was added to include the requirements for GFCI protection of outdoor receptacles at dwellings, but only where there was "direct grade level access to the dwelling and to the receptacle." The 1978 "preprint" on page 31, Proposal 48 was responsible for this.

Section 210-8(b) which addressed GFCI protection for construction sites, was revised to introduce an exception for the basic requirement, allowing what has become known as the "assured equipment grounding system," Exception 2.

This was an attempt to accommodate the fact that there is a vast difference between the construction sites for small projects like one family dwellings and enormous construction sites, some active for a decade, like nuclear generating stations and massive chemical plant complexes.

The justification was that where a well organized program was in effect to monitor the integrity of the grounding conductor at regular intervals, not exceeding 3 months, the system was equivalent to GFCI protection. This was brought to a head by hearings held in Washington by OSHA.

As explained in Chapter 3–Grounding Practices in North America, there is no way, even if the integrity of the grounding conductor were checked every second, that it is equivalent to GFCI protection. GFCIs protect against both Series and Parallel contact, and Grounding only contributes to safety in a Parallel contact condition.

Section 517-13, Ground Fault Protection of Health Care Facilities, was revised to require a second level of equipment ground fault protection, GFP, in the feeders on the load side where GFP is required for the service.

Selectivity was specified to require the GFP closest to the fault to trip first. The 1978 "preprint" on page 304, Proposal 56, led to this change.

Section 517-92, Wet Locations in Health Care Facilities, required GFCI protection if power interruptions could be tolerated. The 1978 "preprint" on page 321, Proposal 115, originally rejected, initiated this change.

Section 550-6(b) was revised to require that in Mobile Homes, all 15 A and 20 A receptacles installed outdoors and in bathrooms shall have GFCI protection. This was previously assumed to be required per a footnote in Section 550-6(a), but this clarified the requirement. The 1978 "preprint" on page 338, shows that Proposal 17 was responsible.

Section 551-7(c) required that receptacles in recreational vehicles, in bathroom facilities and outdoors, have GFCI protection. The 1978 "preprint" on page 347, Proposal 64 was responsible.

Section 551-42 required that all 15 A and 20 A receptacles at recreational vehicle park sites have GFCI protection. The 1978 "preprint" on page 352, Proposal 94, paved the way.

Section 555-3 was revised to add a requirement that at marinas and boatyards, the 15 A and 20 A receptacles other than those supplying shore power to boats, have GFCI protection. The 1978 "preprint" on page 354, Proposal 2, initiated this change.

1981 NEC

Section 210-8(a)(2), GFCI Protection of Receptacles in Garages of Dwelling Units, was revised to add two exceptions to the basic requirement:

> Exception No. 1 exempted receptacles which were not "readily accessible." This meant that the receptacle over the garage door supplying the automatic door operator would not have to be GFCI protected.

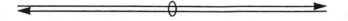

Exception No. 2 exempted receptacles "for appliances occupying dedicated space." This addressed the claim that people should be able to install refrigerators and freezers in their garage, regardless of their condition, and not have food spoil from an undetected tripping of a GFCI.

This is a good example of how things get into the NEC through the consensus process. The proponents of the exception claimed that there were 4 cases where GFCIs tripped in garages, and food was spoiled in refrigeration equipment. No evidence was presented as to the electrical condition of the refrigeration equipment. Further investigation by the author discovered that in one case, the refrigerator was on the same circuit as the garage door operator, and both were controlled by the same wall switch. The user, before leaving for vacation, turned the switch off to prevent inadvertent opening of the garage door. Naturally the food spoiled, but the GFCI got the blame.

Reaching a consensus within any committee frequently requires compromise, and the proponents for GFCI protection went along with the exceptions in order to prevent the "baby from being thrown out with the bath water."

The 1980 document is referred to on the cover as a "preprint," and inside the front cover as the "Technical Committee Report." This was the beginning of the "TCR" terminology. On Page 27 of this document, Proposal 48A initiated the change. The final wording evolved at the second review stage, and is documented in the Technical Committee Documentation, or TCD, on page 12, Proposal 70-22.

Section 600-11, Outdoor Portable Signs. A requirement for GFCI protection of the internal wiring of outdoor portable signs was added, effective Jan. 1, 1982. The 1980 TCR shows on page 330 that Proposal 73A was the source.

Section 680-41(a)(2) added requirements that receptacles located within 20 ft. of spas and hot tubs be GFCI protected. In the 1980 TCR, this is documented on page 317, Proposal 54.

Section 680-62(a), Therapeutic Tubs. GFCI protection was added. Section 680-62(f) required GFCI protection for all receptacles within 5 ft of theraputic tubs. The 1980 TCR, page 318, Proposal 61 was the originator of both requirements.

1984 NEC

In Article 100 a definition was added for "ground fault protection of equipment." The 1983 TCR, page 9, Proposal 1-52 is the originator.

Section 210-7(d) addresses the replacement of receptacles on branch circuits. The basic requirement is that 2–wire receptacles are to be replaced by properly grounded 3–wire grounding–type receptacles. There has been an exception allowing 2–wire receptacles to be used as replacements "where a grounding means does not exist in the enclosure." This was modified to allow a GFCI type receptacle, provided the GFCI receptacle does not supply other receptacles. The 1983 TCR, page 28, Proposal 2-18 is responsible.

Section 210-8(b). Requirements for GFCI protection at construction sites were transferred from this section to Section 305-4. Article 305 covers temporary wiring. A new 210-8(b) was accepted requiring that receptacles in bathrooms of hotels and motels have GFCI protection. The 1983 TCR, page 28, Proposal 2-22 documents these changes.

Section 600-11 was revised, allowing the GFCI protection of portable signs to be in the plug of the supply cord. The 1983 TCR, page 402, Proposal 18-230 was the source.

Section 680-26(b). A requirement was added that the supply to electric motors and controllers of electrically operated swimming pool covers must be protected by GFCIs. The 1983 TCR, page 344, Proposal 20-48 originated this change.

1987 NEC

Section 210-7(d). The exception to the requirement that 2–wire receptacles be replaced with 3–wire, grounding–type receptacles was revised to allow the GFCI receptacle to supply other receptacles, and to prohibit a grounding conductor from being extended to those receptacles. The 1986 TCR, page 64, Proposal 2-57 originated the change.

Section 210-8(a)(4) was revised to require that in dwellings, at least one receptacle in the basement have GFCI protection. The 1986 TCR, page 72, Proposal 2-84 was the source.

Section 210-8(a)(5). This section was added to require that in kitchens of dwellings, all receptacles within 6 ft. of the kitchen sink, above counter tops, shall have GFCI protection. The 1986 TCR, page 67, Proposal 2-69 introduced the requirement. The proposal at the TCR stage did not have the necessary 2/3 affirmative vote. Many comments were received supporting the proposal, including the decisive one, 2-66 on page 43 of the 1986 TCD, and the panel accepted it.

This proposal had a particular significance. Up until this requirement all the GFCI requirements in a dwelling could have been served through one GFCI. As a practical matter this was rarely done because of wiring simplicity and user convenience. Since the NEC required at least two small appliance branch circuits in the kitchen dedicated for the kitchen appliances, with minor exceptions, and because of the required spacing of the receptacles, it was necessary to have at least one additional GFCI in the dwelling. This at least doubled, and probably tripled the GFCI market. Meanwhile, with improved designs and increased volume, the prices of GFCI's came down dramatically, adding to their acceptance.

Section 210-8(a)(6). A new requirement was added that receptacles in boathouses of dwellings must have GFCI protection. The 1986 TCR, page 72, Proposal 2-96 triggered this requirement. The proposal was referred to Panel 20 for action by the correlating committee, and Panel 20 accepted it per the 1986 TCD, page 45, Comment 20-1.

Section 422-8(d)(3). A new requirement was added that the supply cord and internal wiring of portable high pressure spray washing machines be protected by GFCIs. The GFCI was permitted to be an integral part of the plug. The 1986 TCR, page

393, Proposal 10-3 was the initiating proposal.

Section 427-22. A new Section, ' Equipment Protection' was added to Article 427 - Fixed Electric Heating Equipment for Pipelines and Vessels to require that ground- fault protection of equipment be provided for branch circuits supplying electric heating equipment not having a metal covering. The original Proposal No. 12-7 on page 405 of the 1986 TCR was accepted to require GFCI protection. However, Comment 12-2 on page 258 of the 1986 TCD persuaded the panel to change to GFP. The panel was concerned about nuisance tripping of 6ma GFCIs from capacitance- type leakage from the heating cables. The panel stated that its intent is to minimize the possibility of fire and to require GFP as defined in Article 100.

Section 511-10. A new requirement was added to Commercial Garages, Repair, and Storage facilities for GFCI protection of 15 A and 20 A receptacles where electrical automotive diagnostic equipment, electrical hand tools and portable lighting devices are used. The 1986 TCR shows this Proposal as No. 14-72 on page 451.

Section 680-70 was added to require that the circuits supplying hydromassage bathtubs and associated components be GFCI protected. The 1986 TCR traces this change to page 576, Proposal 20-117.

1990 NEC

Section 210-7(d), Exception. The exception to the requirement that 2–wire receptacles be replaced with 3–wire grounding–type receptacles, properly grounded, was revised. A new sentence was added permitting 2–wire receptacles to be replaced by 3–wire receptacles, without grounding connections to the grounding terminal, where the receptacles were supplied through a GFCI type receptacle.

This is a highly–controversial subject, and comes the closest of any previous NE Code controversy in causing people to compare the relative merits of "grounding" versus GFCI protection.

The "pro–grounding only" faction pointed to the historical

acceptance of grounding, and the understanding that peop
have, that where there is a grounding blade and a matchir
grounding opening, there is a good ground.

The "pro-GFCI" faction agreed that where proper groundir
can be accomplished, it should be, but the choice under co:
sideration was whether an unusable 2-wire receptacle shou
be replaced with an old design 2-wire receptacle, or a 3-wir
receptacle without ground, protected by a GFCI. They claime
that the GFCI-protected 3-wire receptacle without ground is f:
safer than the 2-wire receptacle. We note that later issues of tl
NE Code continue to struggle with this subject. The 1989 TCI
page 53, Proposal 2-34 documents this change.

Section 210-8(a)(4) was revised to limit the GFCI requir
ments in basements to "unfinished" basements, avoiding tl
split-level home interpretation issue. Also, GFCI requiremen
were added to receptacles in crawl spaces at or below grad
There were 3 exceptions. The 1989 TCR reference is on page 5
Proposal 2-49.

Section 215-10, Feeder Ground Fault Protection of Equi
ment, was added to require the same GFP for feeders as applie
to Services in Section 230-95. The 1989 TCR, on page 82, Pr
posal 2-180 was the source.

Section 422-8(d)(3)-The requirements for GFCI protection
high pressure spray washing machines was revised to requi:
that the protection be in the plug or within 12" of the plug. Tl
1989 TCR, page 381, Proposal 10-12 triggered the change.

Section 422-24, Cord- and Plug-Connected Appliances Su
ject to Immersion. This new section initiated a whole new gene
ation of electronic safety devices. It requires that portable hydr
massage units and hair dryers be constructed to protect again
electrocution when in the "On" or "Off" position.

This is a good time to tell the reader about one of the mo
valuable tools available to us for indicating where electric
problems need attention. We had access to a national newsp
per clipping service that sent us articles from the whole US
on electric shock and electrical fire accidents. These would hig

light possible problems, and we could dig deeper where desired to get more facts. When I was chairman of Code Panel 2, I made this information available to the panel, as appropriate.

One of the repeat electrocution scenarios was the situation where a hair dryer was left near the edge of a bathtub, plugged in, but turned off. While someone was taking a bath, the hair dryer would somehow fall into the water, killing the bather. I will never forget one particular incident. Twin boys, 2 years old, were in the tub bathing with their mother attending. The phone rang, and the mother left them just for a few minutes to answer the phone. When she returned, both children were dead! A hair dryer had fallen into the tub.

We attempted to alert the public with limited success, that, for gosh sakes, unplug those hair dryers and curling irons around tubs! Compounding the problem were the presumably intelligent adults who insisted on drying their hair while in a shower or tub! Several fatalities have resulted.

To make appliances with heating elements in them electrically safe when submerged is no easy task. New homes have GFCI protection, but there are an awful lot of old homes without GFCIs. In response to this need Leviton Mfg. Co. first invented and introduced a product called an "Immersion Detection Circuit Interrupter" or IDCI. We'll cover them in detail in Chapter 15.

The IDCIs became a viable solution for making the immersed appliance electrically safe, and backed up the NE Code requirement. The 1989 TCR, page 385, Proposal 10-28 initiated this significant new requirement.

Section 550-8(b) was revised to require GFCI protection for all receptacles within 6 ft of any lavatory or sink within mobile homes. The 1989 TCR, page 511, Proposal 19-24 originated this change.

Section 551-41(c)(2) was revised to require GFCI protection for all receptacles within 6 ft of any lavatory or sink in recreational vehicles.

Section 690-5, Solar Voltaic Systems, required that roof–

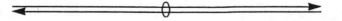

mounted solar photovoltaic arrays on dwellings be provided with "ground-fault protection to reduce fire hazard." The 1989 TCR shows on page 587 that Proposal 3-110 initiated the requirement.

1993 NEC

Section 210-7(d), Replacements. A new paragraph was added to require that GFCI protected receptacles be provided where replacements are made at receptacle outlets that are required to be so protected elsewhere in the Code. The 1992 TCR, page 49, Proposal 2-42 originated this requirement.

Section 210-7(d), Exception. This was revised to require that where 2-wire receptacles are replaced by 3-wire grounding-type receptacles supplied through a GFCI, the receptacle locations shall be marked "GFCI Protected." The 1992 TCR, page 49, Proposal 2-42 originated this requirement.

Section 210-8(a)(5) was revised to include a requirement for GFCI protection for receptacles within 6 ft of wet bar sinks in dwellings. The 1992 TCR, page 51, Proposal 2-56 was responsible for the change.

Section 210-8(b) was expanded to include much more than hotel and motel bathroom GFCI requirements. Section 210-8(b)(1) required GFCI protection for receptacles in bathrooms of commercial, industrial, and all other non-dwelling occupancies. The 1992 TCR, page 59, Proposal 2-117 was the source. Section 210-8(b)(2) required GFCI protection for receptacles on roofs of other-than-dwelling units. The 1992 TCR, page 60, Proposal 2-118 initiated the requirement.

Section 426-53. A new section, "Equipment Protection", was added to Article 426 - Fixed Outdoor Electric Deicing and Snow-Melting Equipment to require that ground- fault protection of equipment be provided for branch circuits supplying fixed outdoor electric deicing and snow- melting equipment. The proposal which initiated this requirement was Proposal 20-53 on page 369 of the 1992 TCR.

Section 600-11 was revised to require that GFCI protection

for outdoor portable signs be in the power supply plug or within 12" of the plug. This ensured that the entire sign would be protected, including the power supply cord. In the 1992 TCR this is on page 523, Proposal 18-93.

Section 620-85 was added to require GFCI protection for receptacles in elevator machine rooms, machinery spaces, pits, and elevator car tops. The 1992 TCR, page 539, Proposal 12-100 initiated the requirement.

Section 680-42 was added, requiring that spas and hot tubs and associated electrical components be GFCI protected, effective Jan. 1, 1994. Previously, just the receptacles for cord– and plug–connected units were required to be GFCI protected. The 1992 TCR, page 569, Proposal 20-149 led to the requirement. The proposal was initially rejected, but later it was accepted as a result of Comment 20-115 on page 607 of the 1992 TCD.

1996 NEC

Beginning with the 1996 NEC, NFPA renamed the document previously named the "Technical Committee Report", or TCR, as the "Report on Proposals", or ROP. The document previously named the "Technical Committee Documentation", or TCD became the "Report on Comments", or ROC.

Section 210-7(d) was revised to require that GFCI receptacles of the grounding type, without an equipment ground connection, where used as replacements for 2–wire receptacles, be marked "No Equipment Ground." Three–wire receptacles without an equipment ground connection, where used as replacements for 2–wire receptacles, and where protected by a GFCI, are required to be marked "GFCI Protected" and "No Equipment Ground." The 1995 "Report on Proposals" (ROP), page 50, Proposal 2-58a, was the source.

Section 210-8, GFCI Protection for Personnel, was completely rewritten. As part of the rewrite the following GFCI requirements were added:

> Section 210-8(a)(2) was revised to add "grade–level portions of unfinished accessory buildings used for

storage or work areas" to the section that requires GFCI protection of receptacles in garages of dwellings. This is in the 1995 ROP, page 59, Proposal 2-105.

Section 210-8(a)(3) was revised to delete the limitation "where there is direct grade level access to the receptacles" from the requirement for GFCI protection of outdoor receptacles at dwellings. Thus, outdoor receptacles at dwellings will be required to have GFCI protection, including those on balconies of high rise construction. It is noted that there is no requirement for receptacles to be installed on such balconies, but if they are installed, they must have GFCI protection.

There is an exception for receptacles that are not readily accessible and are supplied from a dedicated branch circuit for electric snow melting or deicing equipment as covered in Article 426. The 1995 ROP, page 61, Proposal 2-117, was the source.

Section 210-8-(a)(6) was added, extending the GFCI requirements in the kitchens of dwellings to all receptacles "to serve counter top surfaces," instead of limiting the requirement to those within 6 ft of the kitchen sink.

In the rewrite process of Section 210-8, the requirements for GFCI protection of receptacles in boathouses of dwellings, formerly in Section 210-8(a)(6), were deleted, because this is covered by Section 210-8(a)(2). The rewrite of Section 210-8 is in the ROC, Page 30, Comment 2-26.

Section 305-6, Temporary Wiring Ground- Fault Protection for Personnel, was revised to clarify where GFCI protection is required. However, the illusion is maintained that the Assured Equipment Grounding Conductor Program of 305-6(b) provides a comparable level of protection to GFCIs. Protection from only a parallel shock scenario as provided by 305-6(b) cannot be equivalent to the protection from both a series and a parallel shock scenario as provided by 305-6(a). The changes are in the 1995 ROC in Comment 3-140 on page 166.

Section 426-28. Equipment Protection. A new section was added requiring GFP protection for branch circuits supplying fixed outdoor electric deicing and snow - melting equipment. An editorial change transferred the exact wording of Section 426-53 to the new Section 426-28 and deleted Section 426-53. In the 1995 ROP, this is covered by Proposal 20-40 on page 485 and Proposal 20-45 on page 486.

A new Article 552, Park Trailers, was accepted and Section 552-41(c) requires GFCI protection for the receptacles adjacent to a bathroom lavatory, within 6 ft of any lavatory or sink, in the area occupied by a toilet, shower or tub, and on the exterior of the unit. This may be found in the 1995 ROP, page 659, Proposal 19-133a. In the 1995 ROC this is covered in Comment 19-105a on page 433.

Section 620-85 was revised to add "escalator and moving walk wellways" to the elevator and similar areas where receptacles are to have GFCI protection. Section 620-85 was also revised to clarify that the required GFCI protection is to be provided by receptacle–type GFCIs. The ROP, page 695, Proposal 12-79 was the source.

Article 625, Electric Vehicle Charging System Equipment, was added to the NE Code. Section 625-22 specifies that ALL electric vehicle supply equipment shall have GFCI protection. Where cord– and plug–connected electric vehicle supply equipment is used, the GFCI protection shall be in the plug or within 12″ of the plug. The 1995 ROP, page 697, Proposal 12-86, originated the requirement.

Section 680-6(b), Exception No. 1, was revised to require the existing lighting fixtures permitted by the exception to be protected by a GFCI. The 1995 ROP, page 722, Proposal 20-67 was the source.

Section 680-70 was revised to require GFCI protection of all receptacles within 5 ft. of hydromassage tubs.

In the 1995 ROP, this is on page 733, Proposal 20-124.

1999 NEC

Section 210-8(a)(2) was revised to extend the GFCI requirements to accessory buildings that have a floor located at or below grade level not intended as habitable rooms.

This was the result of Proposal 2-60 on page 100 of the 1998 ROP.

Section 210-8(b) was revised to add an exception to the requirement for GFCI protection of receptacles on rooftops of other than dwelling units. The exception accommodates snow-melting and deicing equipment installed in accordance with Article 426.

Proposal 2-96 on page 106 of the 1998 ROP initiated the change . It was revised by Comment 2-45 on page 90 of the 1998 ROC.

Section 210-12, Arc- Fault Circuit- Interrupter Protection. This new section was added to recognize the emergence of an entirely new technology that can detect arcs which do not cause a residual current condition necessary to activate protection such as GFCIs and GFP. It is necessary for such arc fault protection to discriminate between normally harmless arcs such as those created by switching operations and motor arcing, and potential harmful, fire- causing arcs. Section 210-12(a) provides a definition of an arc- fault circuit- interrupter, an AFCI. 210-12(b) adds a requirement for AFCI protection for receptacles in bedrooms of dwelling units, effective 1-1-2002. The panel rationalized that in order for the new technology to be used and to mature, an initial requirement in the NEC was the best way to proceed. Proposal 2-130 on page 115 of the 1998 ROP and related proposals initiated this requirement. Panel action on Comment 2-65 on page 99 of the 1998 ROC finalized the requirement. A new Chapter 16 has been added to this text to discuss the AFCI technology in detail.

Section 215-10, covering GFP for feeders, has been rewritten to correlate it with 230-95 and 240-13. Exceptions were added for continuous industrial processes and fire pumps.

Proposal 2-267 on page 142 of the 1998 ROP initiated the change.

Section 305-6 for GFCI requirements of temporary wiring has been revised as follows:

305-6(a) requires GFCI protection of 125v 30A as well as 15 and 20A receptacles.

305-6(b) requires either GFCI protection or an assured equipment grounding conductor program for receptacles other than 125v, 15A, 20A, or 30A receptacles. This was initiated by Proposal 3-175 on page 394 of the 1998 ROP. The wording was fine-tuned as a result of Comment 3-153 on pages 293 and 294 of the 1998 ROC.

Section 410-58(a), Grounding Poles of grounding- type devices, has been revised to permit the grounding pole of a plug-in GFCI to be of the self- restoring type. Proposal 18-74 on page 646 of the 1998 ROP is responsible for this change.

Section 424-44(g) has been added to require GFCI protection for conductive heated floors of bathrooms, hydromassage bathtub, spa, and hot tub locations. Proposal 20-69 on page 671 of the 1998 ROP triggered the change. Comment 20-74 on page 516 of the 1998 ROC led to the final wording.

Section 426-28 was revised to permit the GFP protection for fixed outdoor electric deicing and snow- melting equipment to be at locations other than at the branch circuit overcurrent device, thus permitting self- contained, receptacle, or other types.

Excepted from the requirement is equipment that employs mineral- insulated, metal- sheathed cable embedded in a noncombustible medium. Proposal 12-7 on page 673 of the 1998 ROP initiated the change, and the panel action on Comment 12-3 on page 518 of the 1998 ROC refined the wording.

Section 427-22 was revised to permit the GFP protection for

electric heat tracing and heating panels to be at locations other than at the branch circuit overcurrent device, thus permitting self- contained, receptacle, or other types. Proposal 12-10 on page 519 of the 1998 ROP is the source of the change. The exception was split into two exceptions as a result of Proposal 12-25 and 12-26 on page 676 of the 1998 ROP and Comment 12-9 on page 519 of the 1998 ROC.

Section 427-27, Voltage Limitations, formerly 427-26, was rewritten to incorporate into the basic requirement the exception permitting the secondary winding of the isolation transformer connected to the pipeline or vessel being heated to have a voltage greater than 30 but not more than 80v where GFCI protection is provided. This originated with Proposal 12-32 and panel proposal 12-32a on pages 677 and 678 of the 1998 ROP. This was fine- tuned by Comment 12-13a and 12-14 on page 520 of the 1998 ROC.

Section 518-3(b) addresses temporary wiring in places of assembly. The section has been rewritten to blend the intent of Exceptions No. 1 and No. 2 into the basic requirements. The panel has maintained the position that the GFCI requirements of Section 305-6 shall not apply to temporary wiring in places of assembly. This originated in Panel Proposal 15-17a 0n page 869 and 870 of the 1998 ROP.

Article 525, Carnivals, Circuses, Fairs, and Similar Events. Section 525-18, which previously stated that the GFCI requirements of Section 305-6 shall not apply to Article 525, has been revised to require in 525-18(a), GFCI protection for "general use" receptacles. The panel maintained the position in 525-18(b), Appliance Receptacles, that receptacles supplying items such as cooking and refrigeration equipment,"which are incompatible with GFCI devices", shall not be required to have GFCI protection. Here we go again. Why do people continue to believe that UL listed appliances are OK when they trip GFCIs? I can just picture the old pop corn appliances, electric cookers, and old portable refrigerators, carted from circus to circus, operating in all kinds of weather, that are deemed to be "incompatible with

GFCI devices". You bet they are. UL lists such devices as having a maximum allowable leakage current of 0.5 ma. If they are tripping GFCIs, they are most likely leaking more than 6.0ma. How does one prevent an "appliance receptacle" from being used as a "general use receptacle"? Revisions were the result of Proposals 15-68 through 15-72 on pages 890 and 891 of the 1998 ROP.

Section 547-9(c), Receptacles in Agricultural Buildings. A new section was added requiring GFCI protection of all 125v, single phase, 15- and 20- ampere general purpose receptacles in areas having an equipotential plane. This originated with Panel Proposal 19-37a on page 912 of the 1998 ROP. The wording was finalized with Comment 19-15 on page 623 of the 1998 ROC.

Section 550-8(g), Mobile Home Pipe Heating Cable Outlet, was revised to require that the outlet be connected to an interior branch circuit other than a small appliance branch circuit and on the load side of a GFCI. This is an attempt to address the problem discussed in Chapter 15, under the Leakage monitoring Receptacle, or LMR. The proposal was 19-57 on page 915 of the 1998 ROP. Panel Comment 19-19a refined the wording.

Section 552-41(d), Park Trailer Pipe Heating Cable Outlet, was added to require that where a pipe heating cable outlet is installed, it shall be installed on an interior branch circuit other that a small appliance branch circuit and on the load side of a GFCI. This is a parallel to the mobile home requirement in Section 550-8(g). It was initiated by Panel Proposal 19-138a on page 935 of the 1998ROP. It was refined by Panel Comment 19-43a on page 628 of the 1998 ROC.

Section 625-2, Definitions. Electric Vehicle Charging System. A definition was added for Personnel Protection System. This was initiated by Proposal 12-90, on page 966 of the 1998 ROP and modified by Comment 12-48 in the 1998 ROC.

Section 625-22, Personnel Protection System, formerly entitled Ground- Fault Protection for Personnel, was revised to describe more accurately the family of personnel protection devices and construction features to provide a level of personnel protection equivalent to the GFCI. This is included in Proposal

12-90, on page 966 of the 1998 ROP. It was modified by Com ment 12-48 on page 666 of the 1998 ROC.

Section 680-57 is a new section with requirements for sign installed in fountains. It requires that all circuits supplying th sign have GFCI protection. This originates with

Proposal 20-201 on page 1012 of the 1998 ROP, as modified b Comment 20-145 on page 691 of the 1998 ROC.

Section 680-62(a), Therapeutic Tub Protection, has been revise to include the outlet supplying packaged and field assembled uni in the GFCI requirement. Exempted are field assembled uni rated greater than 250v or rated three phase.

Section 680-62(b). The present 680-62(a) becomes 680-62(and continues to require GFCI protection of all therapeut equipment.

Section 690-5, Ground-Fault Protection of roof- mounte dc photovoltaic arrays located on dwellings has been adde to require dc ground- fault protection to reduce fire hazard 690-5(a) requires that the GFP device or system shall be capab of detecting a ground fault, interrupting the flow of a fault cu rent, and providing an indication of the fault.

Section 690-6, Alternating – Current Modules has been adde and 690-6(d) permits the use of a single detection device detect only ac ground faults and to disable the array by remo ing ac power to the ac module. The changes in Article 690 we initiated by Proposal 3-217 on page 1019 of the 1998 ROP ar refined by Comment 3-176 on page 695 of the ROC.

A summary of the significant GFCI, GFP, and AFCI requir ments in the NEC and the issue of the NEC where they fir appeared is included in the following table :

Summary of Significant GFCI, GFP, and AFCI Requirements in the NEC

Requirements are for GFCI protection unless otherwise noted.

NEC Issue	NEC Section	Requirements in the NEC
1968	680-4(g)	Swimming pool underwater lighting.
1971	210-7	Construction sites, effective 1-1-74.
	210-22(d)	Outdoors of dwellings, effective 1-1-73.
	215-8	Feeders.
	230-95	GFP.
	680-6	Swimming pool receptacles.
	680-31	Storable pool equipment.
1975	210-8(a)	Became requirements for outdoors of dwelling and added bathrooms of dwellings plus definition of "Bathroom".
	210-8(b)	Became requirements for construction sites.
	680-41(a)	Fountains.
1978	210-8(a)(1)	Garages of Dwellings.
	517-13	Second level of GFP in health care facilities.
	517-92	Wet Locations of health care facilities.
	550-6(b)	Outdoors and bathrooms of mobile homes.
	551-7(c)	Recreational vehicles.
	551-42	Recreational vehicle parks.
	555-3	Marinas and boatyards.
1981	600-11	Internal wiring of outdoor portable signs.
	680-41(a)(2)	Spas and hot tubs.
	680-62(a)	Therapeutic tubs.
1984	210-7(d)	Replacement receptacles
	210-8(b)	Bathrooms of hotels and motels.

NEC Issue	NEC Section	Requirements in the NEC
	305-4	Became requirements for construction sites.
	600-11	Plugs of portable signs.
	680-26(b)	Swimming pool cover motors.
1987	210-8(a)(5)	Kitchens. - Receptacles within 6' of sink.
	210-8(a)(6)	Boathouses of dwellings.
	422-8(d)(3)	High pressure spray washers.
	427-22	GFP for heating of pipelines and vessels.
	511-10	Commercial garages.
	680-70	Hydromassage bathtubs.
1990	215-10	GFP on feeders.
	422-24	Appliances subject to immersion (IDCI)
	550-8(b)	Mobile home sinks.
	551-41(c)(2)	Recreational vehicle sinks.
	690-5	GFP of solar voltaic systems.
1993	210-7(d)	GFCI protection for replacement receptacles.
	210-8(a)(5)	Wet bar sinks in dwellings.
	210-8(b)(1)	Bathrooms of all non-dwelling occupancies.
	426-53	GFP for deicing and snowmelting equipment.
	620-85	Elevator machine rooms and spaces.
1996	210-7(d	Marking of replacement receptacles.
	210-8	Complete Rewrite
	210-8(a)(2)	Accessory buildings of dwellings.
	210-8(a)(3)	All outdoor receptacles including balconies.
	210-8(a)(6)	All receptacles in kitchens "to serve countertop surfaces".
	305-6	Temporary wiring requirements revised.
	426-28	GFP for deicing and snow-melting equipment
	620-85	Escalators and moving walkways.
	625-22	Electric vehicle charging equipment.

NEC Issue	NEC Section	Requirements in the NEC
1999	210-8(a)(2)	GFCI protection for accessory buildings.
	210-8(b)	GFCI exception for snow melting receptacles.
	210-12	AFCI protection.
	215-10	GFP on feeders revised.
	305-6	GFCI protection of temporary wiring.
	410-58(a)	Plug-in GFCI with self- restoring grounding blade.
	424-44(g)	GFCI protection of conductive heated floors.
	426-28	GFP protection of snow- melting equipment.
	427-22	GFP protection of heat tracing and heating panels.
	427-27	Voltage limitations with GFCI protection.
	518-3(b)	No GFCI for exhibition halls.
	525-18(a)	GFCI protection for "general use" receptacles at circuses, etc.
	547-9(c)	GFCI protection in areas having equipotential planes.
	550-8(g)	GFCI for mobile home pipe heating outlet.
	552-41(d)	GFCI for park trailer pipe heating outlet.
	625-2	Electric vehicles – definition of Personnel Protection System.
	625-22	Requirements of personnel protection system.
	680-57	GFCI protection for signs in fountains.
	680-62(a)	GFCI protection for therapeutic tub outlet.
	690-5	GFP for dc photovoltaic arrays on dwellings.
	690-6	GFP for ac photovoltaic arrays.

Chapter 12
Equipment Ground Fault Protection (GFP)

The principle of residual current devices was first applied in the USA by the use of ground fault relaying systems and the application of GFP to large electric generators by electric utilities. When I first went to work for an electric utility in 1946, ground fault relaying had been in use for many years to protect distribution, distribution supply, and transmission lines.

It was in applying this principle to low voltage systems, providing vastly improved electric shock and electrical fire safety, that the use of RCD's really took off, and this is the subject that this text is focused on.

In Chapter 11 we reported that a requirement for GFP was first included in Section 230-95 of the 1971 National Electrical Code. This requirement was in response to a proposal by NEMA with the following supporting comments :

> "Conventional overcurrent devices properly selected in the light of the system of available short–circuit current provide sound equipment protection in the case of normal overloads and low impedance faults such as phase–to–phase and bolted faults. However, high–impedance faults such as arcing–to–ground are more prevalent than bolted faults. In grounded electrical distribution systems, these faults may start phase–to–phase or phase–to–ground. In metal–enclosed equipment, a phase–to–phase fault, if not interrupted immediately by circuit overcurrent protection, will usually quickly become a phase–to–ground fault also. These high–impedance (low–current) faults can cause serious equipment damage before the overcurrent protective device functions.

> This delay greatly increases the risk of injury to
> personnel. There has been an increasing frequency
> of major service equipment damage due to line-to-
> ground arcing faults. Therefore, ground-fault protec-
> tion, which can be set to function on ground-fault
> currents too low to actuate the tripping elements of
> the service overcurrent devices, is needed on low-
> voltage distribution systems.
>
> During the past several years, at least five major cities
> have reported major electrical equipment damage and
> in two instances fatalities have occurred in which
> low-fault currents have persisted for a sufficient time
> to cause very extensive damage."

It has been demonstrated that an arcing ground fault can
occur just about anywhere on an electrical system. A rodent
nesting in a switchgear cubicle, a wire "fishing-in " tape
touching a live terminal, a cable pinched against a sharp
metal edge–that's all it takes to trigger an arcing ground fault
chain reaction.

Arcing ground faults have been known to persist for over an
hour at current levels below those necessary to trip a circuit
breaker or blow a fuse. The energy released can lead to total
destruction of the building service equipment.

For many years the almost universal service voltage for large
buildings in large city centers was 208Y/120 v. Most cities had
a "network" grid to supply the downtown buildings. As the use
of electricity grew, it became more economical in some locations
to install 480Y/277 v "spot networks." An increased incidence
of arcing fault burndowns coincided with the increased use of
480Y/277 v systems.

A ground fault initiated on a 208Y/120 v system is usually
burned clear and is not self-sustaining. A simple analogy can
help explain the reasons why the arc is more likely to persist
at the higher voltage. Studies have shown that an arcing ground
fault can be represented as a spark gap with a spark-over volt-

age of 375 v. Once the arc has been established, the voltage across the conducting gap remains relatively constant at about 140 v. Since an AC voltage is a sine wave, when the voltage passes through zero there is no voltage to sustain the current, and the arc is snuffed out. On a 208Y/120 v system the peak line-to-neutral voltage is 120 x 1.41 = 170 v, considerably below the 375 v spark-over (restrike) voltage necessary to reestablish the arc.

On the other hand, on a 480Y/277 v system, the peak line-to-neutral voltage is 277 x 1.41 = 392 v, which is higher than the 375 v spark-over level, so restrike is likely and the arc persists. Once Section 230-95 of the 1971 NE Code required GFP for the specified parameters, the manufacturers proceeded to make available a wide variety of equipment to meet the requirements.

With the advent of "electronic" circuit breakers, that is, breakers that sensed overcurrents and initiated the tripping action electronically, it became a relatively easy step to include the GFP functions in the same electronics with the addition of either integral or remote current transformers for the residual, or "ground-fault" current sensing.

Electronic circuit breakers made much finer coordination of overcurrent protection devices possible. Thus, the line-side breaker closest to the fault could be set to trip ahead of breakers closer to the supply, minimizing the extent of the power outage.

The same principle could be applied to GFP, setting the breakers closer to the load to be protected to trip at lower residual currents than those nearer to the supply. Of course, the residual current, or ground fault current values at which GFP equipment are set are in the several-ampere range, versus 6-thousandths of one ampere for GFCIs. The principle of operation of both GFP and GFCIs is the same, however.

Let's take a hypothetical example to demonstrate how levels of selectivity can be applied to GFP. Figure 28A shows a one-line diagram of an industrial system which has GFP only on the main circuit breaker. There are two feeders shown, Feeder A supplying a critical industrial process, and Feeder B. A ground fault as shown on Feeder B will cause the main breaker to trip, knocking out the building service, power to Feeders A and B, and power to the critical industrial process.

Figure 28B shows a one–line diagram of the same system with levels of selectivity in GFP added to Feeders A and B. Now a ground fault on Feeder B will trip Breaker B, leaving the main service and Feeder A energized. The actual GFP current settings would be determined by the system designer. In our example, we'll select ground fault trip settings of 40 A for GFP-1, controlling the Main Breaker, 20 A for GFP-2, controlling the Feeder A Breaker, and 5 A for GFP-3, controlling the Feeder B Breaker.

Fig. 29 illustrates the various types of components to make up a GFP system.

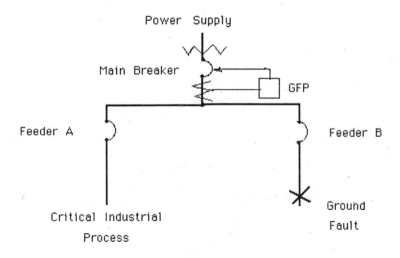

Fig. 28A

GFP Without Selectivity

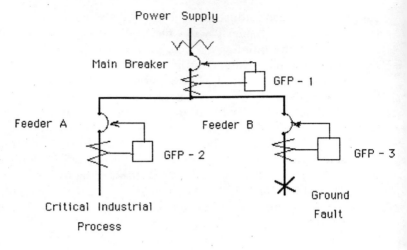

Power Supply

Main Breaker

GFP - 1

Feeder A

Feeder B

GFP - 2

GFP - 3

Critical Industrial
Process

Ground
Fault

Fig. 28B
GFP With Selectivity

Figure 29A
GFP Relay
Courtesy: GE

Fig. 29B
GFP Monitor Panel
Courtesy: GE

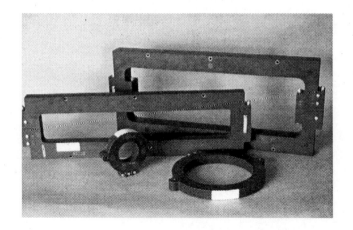

Fig. 29C
GFP Current Transformers
Courtesy: GE

Chapter 13

The Fire Prevention Capabilities of GFCIs and GFP

The capability of residual current devices such as those providing GFP, ground fault protection of equipment, to provide fire protection is well known. What is not so well appreciated is that GFCIs, in addition to providing protection against the harmful effects of electric shock, also prevent certain types of electrical fires from getting started.

To be sure, arcing ground faults are much more likely to continue arcing and to start fires at voltages above the normal 120 v to ground where GFCIs are most usually applied, but arcing can and does occur at the 120 v level.

A good many years ago, when I was traveling the USA on a practically continual basis, participating in meetings of the IAEI, an electrical contractor in St. Louis informed me that GFCIs were "no good." When I inquired why they were "no good" he said that he had to send his men around to loosen the staples in the nonmetallic cables in the homes he was wiring, in order to eliminate the GFCI tripping.

He was using steel, uninsulated staples. I told him he was obviously overdriving the staples and causing current leakage through them. It makes one wonder how many thousand homes there are in the USA with overdriven staples, leaking electricity, wasting energy, and heating up. One thing is sure–without GFCIs protecting those branch circuit conductors where such conditions exist, the conditions will only get worse and could lead to fires.

In this regard, the circuit breaker type GFCI has the advantage over the receptacle type GFCI in that the circuit breaker type protects the entire branch circuit, including the wiring to the first receptacle. The receptacle type protects only the conductors and equipment on the load side of the device. Protection can be optimized with the receptacle type by installing it as the first receptacle on the branch circuit and feeding through it to the other receptacles.

In order to document the fire prevention capabilities of GFCIs, in 1982 I initiated a "fact finding" study with the sponsorship of my employer, conducted by a well–known independent testing laboratory.

The study was divided into two parts: 1) to determine the effectiveness of GFCIs in preventing fires caused by the penetration of nonmetallic cable by metal objects such as nails or staples, and 2) to consider the effectiveness of GFCI–protected lighting circuits as a possible substitute for the thermal protection required for recessed incandescent lighting fixtures. The 1981 NE Code, in Section 410-65(c), was the first issue of the NE Code to require such protection, effective April 1, 1982.

The test set–up for part (1) consisted of a sample "stick built" residential housing wall, using 2x4 wooden studs on 16" centers, with 5/8" gypsum board inside, and outer sheathing for the outside wall. Installations using both No. 14 copper nonmetallic (NM) cable with ground and No. 12 copper NM cable with ground, protected respectively by 15 A and 20 A circuit breakers.

The cables were terminated at a receptacle in a standard outlet box. Thermocouples were installed, and the wall cavity was filled with cellulouse blown insulation to an R–13 value.

An 800 Watt incandescent lamp load was supplied from the receptacle. Thus, at 120 v the current was 6.67 A. A nail was used to penetrate the NM cable, contacting and penetrating the conductors in various combinations. Whenever the nail contacted the black–line conductor and the white–grounded circuit conductor or the uninsulated equipment grounding conductor, the breaker tripped.

Even with only the 6.67 A current, it was possible, on a repetitive basis, to initiate and sustain an arc, which caused a fire. To accomplish this the nail had to sever the white conductor and arc to the equipment grounding conductor, a very rare but possible condition.

Of course, when the circuit breaker was replaced with a GFCI type circuit breaker, it was impossible to create an arc. The GFCI would always trip when a ground–fault of 6 mA or more occurred.

The test set–up for part (2) consisted of a 12 ft by 8 ft simulated attic floor with a roof that gave 5 ft maximum head room. The attic floor consisted of 5/8 " gypsum board nailed to the bottom of 2x6 wood joists. The attic floor formed the ceiling for a room below. The test lighting fixtures were installed in this ceiling. Ten tests were conducted, nine using armored cable and one using NM cable.

Thermocouples were installed and the attic floor space was filled with cellulouse blown insulation to a value of R–19, except for one fixture, which had craft paper–backed fiberglass insulation 6" thick, to an R–19 value. Four types of recessed incandescent fixtures were installed, and they were purposely over–lamped to create severe conditions.

Tests were conducted using standard circuit breakers, standard GFCI type circuit breakers, and GFCI breakers that had been modified to allow measurement of the actual ground fault current. It was my anticipation that it could be shown that enough leakage current would occur to trip a GFCI, way before there was enough temperature rise necessary to start a fire.

The tests did not support this hypothesis. All the fixtures had ceramic lamp sockets, and there was no leakage current measured over 1 mA. There were cases where wire insulation melted, causing enough leakage current to trip GFCIs, but sometimes there was a fire without GFCI tripping. The thermal protection required by NEC Section 410-65(c) is well supported. I guess I'll take a tip from Tom Edison, and note that the test was a success by showing us something that won't work.

Any potential electrical fire condition that causes electrical leakage current to preceed dangerous temperatures, can be alleviated by the application of residual current devices.

Many electrical fires reported in the NFPA Fire Journal could have been prevented if the electrical circuit involved in the fire had been protected by a GFCI. Two documented examples are worth noting :

A nail driven through an electrical cable in a building in New York State caused a fire resulting in a million dollars in damage.

The fire at the MGM Grand Hotel in Las Vegas, which claimed 85 lives, injured about 600 people, and cost more than 30 million dollars in damages, not to mention litigation costs, was caused by a ground fault between an ungrounded circuit conductor and a flexible metal conduit.

It is noted that "overcurrent protection" and " grounding" were present in both of these cases. The NE Code has required some degree of overcurrent protection since its inception. Thus it is logical to conclude that all the buildings experiencing fires of electrical origin in this century have had overcurrent protection. Since the fires occurred, it follows that the overcurrent protection did not prevent the fires.

Some of the fires reported as of electrical origin were no doubt erroneously reported. We have all heard of the cases where building fires were reported as electrical when no electricity was connected to the building. Some electrically–caused fires were beyond the capability of circuit protection to prevent. However, SOME FIRES ARE CAUSED BY ELECTRICITY and could have been prevented by available technology.

A review of the NFPA Fire Journal over the last 5 years for which data are available show "Electrical Distribution Equipment Arcing and Overload" to be a continuing cause of building fires. The annual stastistics are significant–several hundred people killed, several thousand people injured, and hundreds of millions of dollars in fire loss–not to mention the burgeoning legal costs!

When it comes to ground fault protection for equipment or GFP, the big driving force for requiring installations are in the NE Code, Section 230-95 for Services, Section 215-10 for Feeders, and Section 517-17 for Health Care Facilities. GFP systems, which we covered in Chapter 12, can prevent fires that could be caused by arcing ground faults.

A new generation of circuit breaker is appearing which includes the GFP capability, with a trip value in the 30 mA range. These are sometimes referred to as "fire breakers." In Austria "fire breakers" with 100 mA trip levels are required in agricultural buildings.

Now we have the newest kid on the block in our electrical fire prevention capabilities – the arc- fault circuit- interrupter, or AFCI. These devices are covered in detail in Chapter 16.

Chapter 14

The Hair Dryer, etc. and the Proliferation of Protective Device Types

We reported in Chapter 11 how the 1990 NE Code added a requirement in Section 422-24 that cord– and plug–connected portable free standing hydromassage units and hand held hair dryers be constructed to provide protection for personnel against electrocution when immersed while in the "On" or "Off" position.

This requirement is a good example of how the NE Code process embraces new safety concepts. First, there is a safety need which cries out for a solution. The abnormally high fatality rate for people associated with grooming and similar appliances around water, rang the warning bell. These appliances were listed appliances, being used in buildings wired according to the NEC, and yet a lot of people were being killed every year by them.

Investigations into the accidents showed that people were leaving hair dryers, hair curlers, etc. plugged in and placed around the edge of bathtubs on a continuing basis. People did not realize that even though the appliances were turned OFF, if they fell into the bath water they could prove lethal.

From the newspaper articles on the accidents, you could practically write the same article and fill in the blanks. Either the victim was using the hair dryer while bathing, or the appliance was pulled or fell into the tub:

> San Jose, CA–"A 23 year old radio station marketing consultant was killed when his electric hair dryer fell into the tub as he was taking a bath."

> Canton, OH–"An eight–year–old Canton boy was electrocuted and his ten–year–old brother was injured

Monday night when one of them apparently pulled an electric dryer into a bathtub in which they were bathing."

Mansfield, OH–"A 27 year old Mansfield woman died of an electric shock yesterday morning in her home while using an electric hair dryer in her bathtub."

Evansville, IN–"A 9 year old Evansville youth was electrocuted Saturday while apparently drying his hair in the bathtub."

North Park, CA–"A 4 year old North Park girl was electrocuted Monday when a hair dryer fell into the bathtub."

Oakwood, OH–"A 17 year old Oakwood girl, who died Tuesday night a short time after she was found unconcious in her bathtub, died from accidental electrocution and was found in the tub with a 'live' hair dryer in the water."

Mattydale, NY–" A 6 year old boy died Sunday night after suffering a severe electrical shock while taking a bath at his home. He stepped out of the tub, got a nearby hair dryer, stepped back into the water, and turned it on."

Quechee, NH–" A 9 year old Quechee girl died Monday night after being electrocuted in a bathtub accident. Another 16 month old child was in the Intensive Care Unit...a hair dryer either fell or was pulled into the bathtub."

The listing authorities reviewed their standards and revised them to require the appliances to be safe when submerged while in the OFF position. Until then, if the appliance had a single pole ON–OFF switch, normally required to be in the live conductor, unless the 2–wire plug was polarized, there was a 50/50 chance that the interior wiring would always be energized, when the appliance was plugged in. Proper polarization and insulation of the supply terminals added to safety.

However, this did not address the problem of the appliance being turned ON when immersed. This is much more difficult to resolve because appliances with high power heating elements are difficult to make watertight.

GFCIs protecting bathroom receptacles in newer homes or retrofitted homes provide the necessary safety when they are installed. Unfortunately, many homes were constructed before the GFCI requirements and have no GFCI protection.

Manufacturers studied the problem and one manufacturer, Leviton Mfg. Co., invented the IDCI or Immersion Detection Circuit Interrupter, to give protection from immersed appliances. This device is described in detail in Chapter 15.

Now we had the old "chicken vs. egg" dilemma which faces any new product whose only function is increased safety over the status quo. There is a natural resistance by manufacturers to add cost to their products. If the new device is not required, even if a manufacturer were inclined to add it, he is concerned that his competitors will not add the item to their products and will have a cost advantage. Until seat belts were mandated, how many cars had them?

Thus, the answer is to mandate worthy safety features. When a new requirement is added to the NEC, the respective UL standards are revised to satisfy the new NEC requirement. Then all manufacturers have a level playing field.

The burden of convincing the NEC panel, in this case Panel 10, that a viable product existed to make the requirement practical, fell to the company which invented the IDCI, Leviton Mfg. Co.

This was accomplished by demonstrations of products actually being submerged, together with seminars on the products. The federal Consumer Product Safety Commission, CPSC, added its support.

Since the introduction of the IDCI, there has been an accelerated availability of new safety products geared specifically to improving the safety of specific appliances, and built into the appliances.

Chapter 15

Present Variations: GFCI, GFP, ALCI, ELCI, IDCI, and LMR

GFCI

The ground fault circuit interrupter or GFCI is presently listed by UL in two "classes." The Class A GFCI must trip at 6 mA and must not trip below 4 mA. Its tripping curve must be at least as fast as the following curve within the current limits of from 6 mA to 264 mA: $T = (^{20}/_I)^{1.43}$ where T is time in seconds and I is tripping current in milliamperes. The curve is plotted in Fig. 12 and Fig. 14.

For all practical purposes this IS the GFCI. Of the 200 million or so GFCI's in use, all but a few thousand at the most, are Class A.

The second class of GFCI is the Class B, which is limited for use only with underwater swimming pool lighting fixtures. Its use is limited to old swimming pools installed before the 1965 NE Code, which have high leakage lighting fixtures. Its tripping curve must be at least as fast as the following curve within the current limits of from 20 mA to 1,056 mA: $T = (^{80}/_I)^{1.43}$ where T is in seconds and I is in mA. For practical purposes, this device is a Dodo bird.

Where high standing or inherent current leakage make the usage of Class A GFCIs impractical, one approach would be to create other classes C, D, E, etc. I mentioned the conflict between the high leakage allowed for listed personal computers, office equipment, and lab equipment in Chapter 7 and the required 6 mA trip value of Class A GFCIs. However, it appears that the 6 mA Class A GFCI will remain as THE people protector, and that we are unlikely to see additional Classes created. The trend is to give other similar devices completely new names like ALCI, ELCI, etc.

GFP

Ground fault protection of equipment will continue to proliferate with a somewhat hazy line between GFP, GFCI, and ELCI. After all, except for the minimum trip value, if the GFP is set down to a minimum trip setting of 20 or 30 mA, it does provide a high degree of personnel protection against ventricular fibrillation. Likewise, a GFCI will protect equipment from destructive arcing ground faults.

ALCI

The appliance leakage circuit interrupter or ALCI, was created to fill the need for a personnel protection device that could be simpler, less expensive than a GFCI, and more readily mated to the electrical parameters of specific appliances, and built into the specific appliances. The tripping characteristics are identical to a Class A GFCI, but the double–grounded neutral protection feature is not required.

A good application for an ALCI is its use in a beverage can dispensing machine. These machines are often installed in wet locations where normal GFCI protection may not exist. There have been electrocutions where customers, sometimes barefoot in wet bathing suits, have been shocked by dispensing machines. The ALCI is sometimes configured like a blank–faced GFCI receptacle. It also is used in a plug configuration to supply a personnel grooming appliance, such as a hair dryer or hair curler. It has an advantage for use on smaller volume appliances versus the IDCI in that it is a true RCD and does not require a probe wire designed into the appliance, as the IDCI does. Fig. 30 shows an ALCI configured for panel mounting.

ELCI

The equipment leakage circuit interrupter is like an ALCI, except that the minimum trip level is set above the 6 mA value specified for a Class A GFCI or ALCI. Therefore it is not considered a "people protector." It does provide a degree of personnel protection, however. Remember, the laws of physics determine how much current goes through you, not the setting of a protective device. With normally expected human body impedances in the range of 500 to 1000 ohms, the current through a person at 120 v would be in the range of 240 to 120 mA, and a device set at less than those values would trip. Like the ALCI, the ELCI does not have a double-grounded neutral detection feature, but it is a true RCD. It can be set for any trip value, but 20 mA and 30 mA trip settings are more common.

Here we enter a grey area on terminology. Is a circuit breaker RCD configured without double-grounded neutral protection and with a 30 mA residual current trip value an ELCI or a GFP breaker? Take your pick.

Fig. 31 shows a cord connected ELCI with a 27mA trip level for use with heating tapes.

IDCI

The immersion detection circuit interrupter or IDCI, is shown in Fig. 32 in plug form, in a schematic diagram. The IDCI is not a conventional "residual current device." A probe wire is built into the appliance being protected to ensure that if the appliance is immersed in water, current will flow from whichever of the two line conductors is energized, to the probe. This current flow causes solid state components in the plug to conduct, tripping open a 2-pole relay in the plug. The more expensive IDCI's have a reset button in the plug. There are also "one-time" trip devices.

The IDCI probe location and design must be part of the appliance design. Although the IDCI might be the best choice for newly designed, high volume products, the testing and listing process can be complicated and expensive. The ALCI type of plug is more readily fitted to existing designs, requires no probe, and is easier to test.

LMR

The leakage monitoring receptacle or LMR, is similar to an RCD receptacle except that it has the capability to notify the user audibly and/or visually that a given level of leakage current has been reached. It can be designed to trip at some higher level of leakage current.

At this point in time, the LMR is a product waiting for a specific need to define its operational parameters. The product is essentially an ELCI receptacle with alarm capabilities. The LMR was conceived in response to an electrical fire problem under mobile homes. Mobile homes in cold climates often have heating cables wrapped around the water pipes to keep them from freezing. The heating cables are usually supplied from receptacles under the mobile homes.

The heat tapes are frequently installed under thermal insulation, close to the underfloor of the mobile home. Overheating of the heat tape is a possibility. If a fault were to develop in the heat tape, a fire could start, destroying the mobile home. Several mobile home fires each year are attributed to this. However, if the heat tape has an equipment grounding conductor, either as an individual conductor or as a metal sheath, and if the heat tape receptacle is protected by a GFCI, the GFCI trips at 6 mA before the leakage current can become high enough to start a fire.

The dilemma is that the tripping of a conventional GFCI for any reason could go undetected, and this would allow the water pipes to freeze. The LMR addresses this problem. It is technically possible to configure the LMR to have it transmit a warning signal in the form of a flashing light, noise alarm, or both at one leakage current value, say 30 mA, and to trip off the power at another value, say 100 mA. The actual warning and trip levels would be determined by the product listing process. The capability exists, but until a clearly defined requirement exists for the use of the LMR, it probably will not go into production.

Figure 30
ALCI Configured For Panel Mounting
Courtesy: Leviton Mfg. Co.

Figure 31

Cord Connected ELCI with 27mA
Trip Level For Use With Heating Tapes

Courtesy: Leviton Mfg. Co.

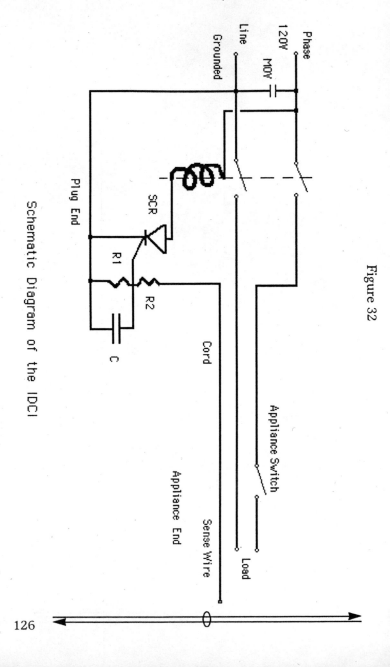

Figure 32

Schematic Diagram of the IDCI

Chapter 16
The Arc-Fault
Circuit-Interrupter

Up until this new chapter in the text, the electrical safety measures discussed in the text were centered around Overcurrent Protection, Grounding, and the many variations in protection based on the principle of the residual current device.

In Chapter 13, The Fire Prevention Capabilities of GFCIs and GFP, we discussed arcing ground faults and how GFCIs and GFP can provide protection from fires initiated by such faults. We also discussed a fact- finding study we originated to investigate the effectiveness of GFCIs in preventing fires caused by the penetration of nonmetallic cable by metal objects such as nails or staples.

The results of the study supported the position that GFCIs will provide a high degree of protection against fires caused by arcing faults, as long as there is a source of ground present to cause the arcing faults to become arcing GROUND faults. It should be kept in mind that the current of an arcing ground fault will likely be too low to activate a standard overcurrent device. Suppose there is no ground path necessary to create the residual current condition to activate a GFCI or GFP device? Until the creation of the AFCI, we had no answer, and our attention was focused on doing things to minimize the likelihood of such conditions occurring. Remember, before we had GFCIs and GFP, we did all we could through equipment grounding and bonding, to attempt to make sure that ground faults would become high level ground faults, so the overcurrent device would clear the fault.

In spite of our efforts to prevent arcing faults, they continue to happen. There are millions of old homes built before there were requirements for equipment grounding conductors. We can still find lots of knob and tube wiring, NM cable, and old BX cable without an equipment grounding conductor. Two- wire extension cords are rampant, including cords used illegally as permanent wiring extensions. Two- wire lighting fixtures, both permanently installed and cord- and plug - connected, exist in the millions.

Last, but not least, abuse of wiring occurs in new as well as in old homes, and the wiring system is vulnerable to abuse for the life of the home.

According to the latest information from the Consumer Product Safety Commission, arcing and sparking in home wiring are associated with an average of over 41,000 home fires annually. These fires annually result in over 350 fatalities, over 1,400 injuries, and over $680 million in property damage.

Fires in extension cord sets and power supply cords of home electronic appliances such as TVs and stereos caused the Electronics Industry Association, representing the electronic appliance manufacturers, to investigate these problems. Their investigations led to NEC Proposal 2-213 on page 76 of the 1995 ROP. This proposal, although rejected by Panel 2, focused industry attention on the problems and ultimately contributed to the creation of the AFCI.

The proposal attempted to require overcurrent protection which would have lower instantaneous activation levels in order to limit the "let- through" current. While this is technically feasible, it should be kept in mind that the lower the instantaneous trip level, the greater the likelihood of unwanted tripping from normal current surges, such as motor starting current. For example, the new generation of more energy - efficient motors have a higher starting current than the older motors, adding to the likelihood of "nuisance tripping". Also, electronic devices of many kinds have proliferated throughout our homes, and these devices are often associated with the creation of transient current spikes.

Panel 2, in its rejection statement commented:

"A more complete analysis of actual cord problems and alternate solutions such as other cord constructions, supplemental overcurrent protection, and electronic sensing is needed."

Up until then, the industry had pretty much accepted the position that such cords were protected by a maximum of 20A overcurrent protection on the branch circuit. If a lower level of protection was desired, fused plugs with smaller rated fuses were available for use. Way back in the 1968 NEC, the oldest one I have handy, we read in Section 240-5(a), Exception No. 3, that No. 18 cord and larger "shall be considered as protected by 20A overcurrent devices". This hasn't changed much through the years. The 1999 NEC, in Section 240-3, permits tinsel cord and No. 18 appliance and portable lamp cord to be supplied from a 20A branch circuit. Section 240-3(g)(3), states that extension cord sets rated No. 16 or larger shall be permitted to be supplied by 20A branch circuits. Extension cord sets are now required to be no smaller than No. 16. However, for decades before this new requirement, two- wire extension cord sets with No. 18 conductor and three 15A outlets were sold by the tens of millions, and many are still in use.

It is important to stress here that regular overcurrent devices have been doing exactly what they were intended to do since their inception. Referring again to the 1968 NEC, Section 240-2 reads as follows:

"Purpose of Overcurrent Protection. Overcurrent protection for conductors and equipment is provided for the purpose of opening the electric circuit if the current reaches a value which will cause an excessive or dangerous temperature in the conductor or conductor insulation."

The protection provided by such devices as GFCIs, GFP, and AFCIs is protection IN ADDITION to that provided by the overcurrent device.

Further discussion of the issue posed by the EIA Proposal 2-213 is covered in EIA Comment 2-94 on page 40 of the 1995 ROC. The Panel Statement rejecting Comment 2-94 adds further information on the technical problems. During the past decade, several manufacturers had been working in their R & D facilities on solutions to the arcing problem. Before a solution could be found, it was necessary to learn a great deal more about the technical details of the problem.

ANATOMY OF THE 120 VOLT ARC :

We discussed in Chapter 12, Equipment Ground Fault Protection, how an arc at 277v to ground is more likely to be sustained until the source power is disconnected, as compared with an arc at 120v. Arcs at 120v, although less likely to be sustained and more likely to burn themselves clear, can sputter and shoot off sparks. They can create a carbonized path which could reinitiate the arc.

Temperatures in arcs can range into the 10,000 degree C range, a veritable arc welder. Arc faults can be categorized into three types : Series Arc Faults, Parallel Arc Faults, and Arc Faults to Ground. Each type of arc fault has its own characteristics.

SERIES ARC FAULTS :

A broken conductor and corrosion or loose connections at terminals are the most likely causes of this type of arc fault. In Chapter 13, The Fire Prevention Capabilities of GFCIs and GFP, we discussed the fact- finding study we conducted to determine the fire prevention capabilities of GFCIs where NM cable was penetrated by metal objects.

We were not able to sustain a true series arc fault in this test. When we initiated series arc faults, they soon became sustained arc faults to ground.

With just a circuit breaker for protection, we were able on a repetitive basis to create an arc to the equipment grounding conductor at 120v by breaking the grounded circuit conductor and bridging the gap with a nail with a current of 6.67A. In our tests, any series arcing at the nail was not sufficient to start a fire. NM cable has the equipment grounding conductor between the two circuit conductors, and the arcing to the equipment grounding conductor did start a fire. Of the three types of arc faults, the series arc fault is the least likely to occur without leading to one of the other two types. It is also the most difficult to protect against without initiating unnecessary tripping.

PARALLEL ARC FAULTS :

In a parallel arc fault, the arc fault occurs between the energized conductor and the grounded circuit conductor, on 120v circuits. All two- wire installations are vulnerable to this type of arc fault, as well as three- wire circuits. A sharp object, abrasion, insulation deterioration, excessive flexing, excessive heat, and moisture are the more common originators of this type of arc fault.

One of the early electrical fire incidences which led the EIA to become concerned about this arc fault problem was a parallel arc fault caused in a two- wire cord which had been penetrated by the metal leg of a piece of furniture resting on the cord. The fault caused a fire resulting in considerable damage. Since the cord was supplying a TV set, the TV manufacturer was drawn into the litigation.

Two-wire heat tapes are a vulnerable type of product for this type of arc fault. This is why the NEC now requires in Section 426-27 that a grounding means, such as copper braid, metal sheath, or other approved means be provided as part of the heated section of the cable and why Section 426-28 requires GFP protection.

ARC FAULTS TO GROUND :

What is meant here is an arc fault not to the grounded circuit conductor, which is grounded at the supply, but an arc fault to the equipment grounding conductor or to grounded metal surfaces. Once an arc fault has gone to the equipment grounding conductor or to grounded metal, it is much easier to detect and to disconnect the supply. This creates a residual current condition, and both GFCIs and GFP provide this protection.

THE CREATION OF THE AFCI :

One of the interesting facts that evolved from the investigative work done on the arc fault problem by the EIA, CPSC, UL,

NEMA, and the manufacturers individually, was how low the available fault currents at receptacle outlets in dwellings actually are.

When I was active in the circuit breaker manufacturing industry, the standard interrupting rating of listed 15 and 20A molded-case circuit breakers ratcheted up from 5,000A to 10,000A. Then 22,000A rated breakers became available as an option. The focus was on the concern that available fault currents were going up as a result of the electric utility trend towards going to lower loss transformers with lower impedances, resulting in higher available fault currents in buildings.

While receptacles installed right at service entrances of dwellings and in commercial and industrial buildings may experience much higher fault currents, the following chart, Figure 33, shows that the available short circuit currents in dwellings are really quite low. The data are from a UL fact-finding report, "An Evaluation of Branch– Circuit Circuit Breaker Instantaneous Trip Levels", conducted for EIA, dated 10-25-93.

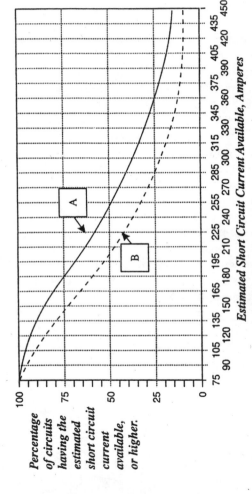

Estimated Short Circuit Current Available, Amperes

Courtesy : Cutler – Hammer

Distribution of Available Fault Currents

Curve A: 15A household receptacles

Curve B: At the end of six feet of #18 appliance wire plugged into those receptacles.

Figure # 33

Note how low the available short circuit currents are at 15A household receptacles. This reduction is due to the impedance of the installed wiring. Similarly, the fault currents are quite a bit lower with 6 ft. of No. 18 cord added.

The vast majority of currents are well below 450A RMS, and half of them are below 250A RMS. We stress the Route Mean Square (RMS) to point out that an arc fault is an intermittent condition, whereas a circuit breaker, when the current is below the instantaneous trip level, responds to RMS values. Thus, an intermittent arc, which might have a peak value of 70A, could actually have an RMS value of about 5A, below the tripping value of a 15A circuit breaker.

Herein lies the need for the AFCI. Present circuit breakers have a long successful history protecting conductors and equipment from overcurrents, but they were not specifically designed to protect against the unique arc fault conditions created in cut or otherwise damaged cords and cables.

In order to determine where in a dwelling arc faults are likely to occur, let's look at data provided by a recent CPSC study and data from a major insurance company.

To aid in the analysis, the dwellings have been divided into four zones. Zone 0 covers the building service and metering. Zone 1 covers the service panelboard or loadcenter, feeders and branch circuits, receptacle outlets, and lighting outlets. Zone 2 covers extension cord sets and power supply cords for lamps, computers, and appliances. Zone 3 covers the utilization equipment. Figure 34 illustrates these zones :

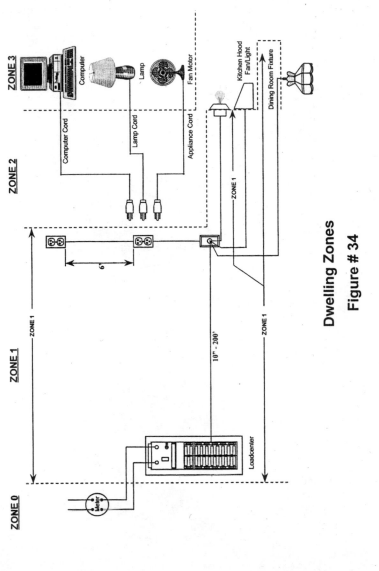

Courtesy : Cutler – Hammer

Dwelling Zones
Figure # 34

135

Two studies of electrical fires in homes, one by a major insurance company covering 690 homes, and one by CPSC covering 41,000 homes, reveal interesting information.

Figure 35 illustrates the data:

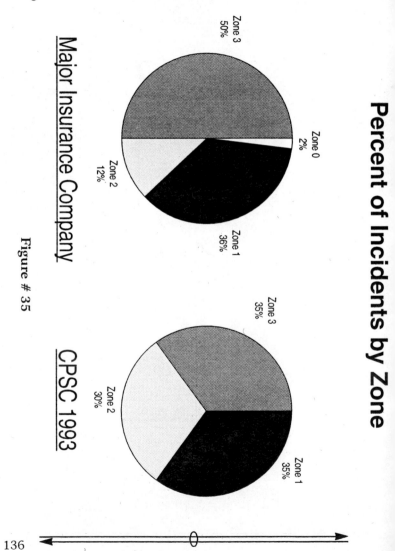

Percent of Incidents by Zone

Major Insurance Company

Zone 3 50%

Zone 0 2%

Zone 2 12%

Zone 1 36%

CPSC 1993

Zone 3 35%

Zone 2 30%

Zone 1 35%

Figure # 35

Zone 0 had 2% in the insurance study and 0% in the CPSC study. Zone 1 had 36% in the insurance study and 35% in the CPSC study. These statistics are remarkably close. Zone 2 and Zone 3 were 12% and 50% in the insurance study as contrasted with 30% and 35% in the CPSC study. One might attribute the difference to some degree with the vagueness of definition and interpretation between the two zones. The sum of these two zones, 62% for the insurance study and 65% for the CPSC study correlates well.

The electrical distribution equipment manufacturers, particularly the circuit breaker and wiring device manufacturers, responded to satisfy the need for better arc fault protection devices. As of this writing, it is reported that there are well over 100 US patents issued on inventions to address arc fault detection and interruption.

During this evolutionary period, several segments of the electrical industry were working together to come up with solutions. UL worked with the NEMA electrical manufacturers to develop a new UL standard for Arc-Fault Circuit-Interrupters, UL 1699, first issued on Feb. 26, 1999. The scope of the standard covers branch/feeder, combination, outlet circuit, portable, and cord type AFCIs intended for use in dwelling units. The devices are intended to mitigate the effects of arcing faults that may pose a risk of fire ignition under certain conditions if the arcing persists. The devices have a maximum rating of 20A and are intended for use in 120v ac, 60-HZ circuits. The devices are not intended to detect growing conditions.

An AFCI that is also intended to perform other functions, such as overcurrent protection, GFCI protection, GFP protection, surge suppression, and any similar functions, or any combination of functions, must comply with the requirements of the applicable other standards, if it is to be listed for those additional functions. Frequently, AFCIs use the residual current principle, in the form of a 30ma GFP-like circuit, to assist in the performance of the device, but the AFCI is not dual-listed as a GFP.

Incidentally, the July/August 2000 issue of the IAEI magazine has excellent articles on AFCIs by Cutler-Hammer, Square D, and UL. The September/October 2000 issue has another excellent article by Pass & Seymour/Legrand. Check them out. To assist in an understanding of the terminology, we have added a number of definitions relative to AFCIs in Chapter 18, Definitions of Terms.

The Branch/Feeder AFCI

This is the one most likely to contain the circuit breaker function as well as the AFCI function. It is possible to be configured as a separate "black box" type at the panelboard. It is more likely to be an interchangeable dual- rated circuit breaker and AFCI , like the GFCI circuit breaker. In fact, devices with circuit breaker, AFCI, and GFCI functions included in one device are already on the market. The devices have two test buttons, one for the AFCI function, and one for the GFCI function.

We discussed series arc faults, parallel arc faults, and arc faults to ground.

UL has tests for these, and a summary of the tests is shown in Figure 36.

Tests	Branch/ feeder AFCI	Combination AFCI	Outlet circuit AFCI		Portable AFCI	Cord AFCI
			With feed	Without feed		
Carbonized path arc ignition test						
NM-B insulation cut	X	X				
Carbonized path arc interruption test						
SPT-2 insulation cut	X	X				
NM-B insulation cut	X	X				
Carbonized path arc clearing time test						
SPT-2 insulation cut		X	X	X	X	X
Point contact arc test						
SPT-2 insulation cut	X	X	X	X	X	X
NM-B insulation cut	X	X				
Unwanted tripping tests						
Load condition I – inrush current	X	X	X	X	X	X
Load condition II – normal operation arcing						
conditions a – c	X	X	X	X	X	X
conditions d – e	X	X	X			
Load condition III – non-sinusoidal waveform	X	X	X	X	X	X
Load condition IV – cross talk	X	X	X			
Load condition V – multiple load	X	X	X	X	X	
Load condition VI – lamp burnout	X	X	X			
Operation inhibition						
Masking	X	X	X	X	X	X
EMI filter	X	X	X	X	X	
Line impedance	X	X	X			

Figure 36

Arc Fault Detection Tests Table

Courtesy : Underwriters Laboratories Inc.

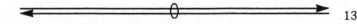

Examination of the table shows that there is a great deal of commonality in the test requirements for the Branch/Feeder, Combination, and Outlet types of AFCIs. Thus they must all recognize parallel(line to neutral) arcing faults in circuits with available short- circuit currents of 75A or above. The main difference between the branch/ feeder requirements and outlet requirements is that the branch/ feeder device is tested for parallel faults in both the installed wire (Type NM-B) and in commonly used two- wire extension and appliance cord (Type SPT-2).

The outlet device, however, is tested solely with Type SPT-2 cord. There is also some commonality in the test requirements for responding to series arcs associated with a break in a line or neutral conductor. For Branch/ Feeder AFCIs the tests are performed at current levels of 5 amps and above on type NM-B cable. The passing criterion is that cotton installed for the test above the break point must not ignite. For Outlet Type AFCIs, the tests are performed at the same current levels with Type SPT-2 cord, and the time for arc extinction must be less than specified arc- test clearing times.

Combination AFCI

Only the Combination AFCI is required to meet all the tests. As of this writing, it is understood that one manufacturer, Pass & Seymour/Legrand, has a device UL listed to this portion of UL 1699.

Outlet Circuit AFCI

This device may also meet the requirements of a receptacle as well as an AFCI. This AFCI is likely to evolve as a dual – rated AFCI and GFCI receptacle, with test buttons for both functions.

Portable AFCI

The portable AFCI has the same additional requirement as the portable GFCI, that it shall provide protection in the event that the grounded conductor becomes open circuited. One way of accomplishing this is with a normally- open relay on the line side. This would make it in the eyes of our European friends, "independent of line power".

Cord AFCI

There is more than one way to skin a cat, although I confess I haven't tried. Technology Research Corp. introduced the " Fire Shield" cord set, a residual current device on the line side of a metallic shielded cord. Any damage to the cord will result in a residual current condition, which will trip the device. It is available as a listed extension cord set and as a recognized component in a power supply cord configuration.

Although this is not a true AFCI, as long as all of the cord on the load side of the device is of the shielded type, the device will certainly protect against arc faults on the cord, as well as residual current conditions at the load end. Figure 37 shows a Fire Shield of the cord set type.

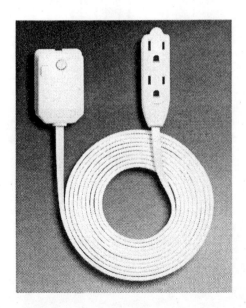

Figure 37
Fire Shield
Cord Set

Courtesy:
Technology Research Corp.

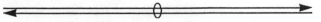

A true Cord AFCI would have to pass the tests listed in Table 50.2 of UL 1699, which includes both series and parallel arc fault tests on SPT-2, unshielded, ungrounded cord.

We are fortunate in receiving the cooperation of Cutler-Hammer in describing one manufacturer's solution to the arc fault problem, so most of our discussion will relate to the Cutler-Hammer approach. Other approaches may differ somewhat, but all UL listed AFCIs will meet the requirements of UL 1699.

Figure 38 shows a typical current waveform when a steel blade cuts through a two conductor, No. 16 SPT-2 cord, protected by a standard 20 A circuit breaker. Notice how erratic the wave form is and that the RMS value is very low.

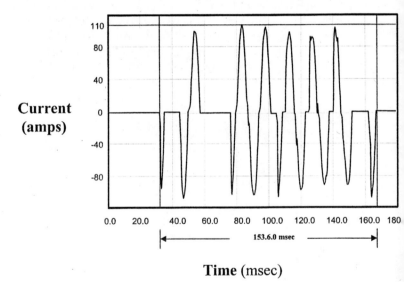

Time (msec)

Figure 38

Typical Current Wave Forms for Blade Cutting No. 16 Cord.
Available Short Circuit Current = 100A

Courtesy: Cutler - Hammer

This sputtering and arcing, possibly shooting off sparks, will likely continue erratically until the supply is disconnected.

Figure 39 shows the same test when the supply is protected by a 20A AFCI circuit breaker.

The supply is disconnected after a minimum of arcing.

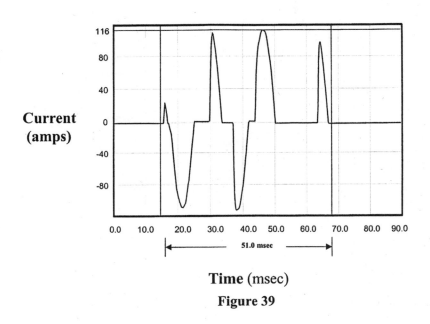

Time (msec)

Figure 39

Wave Forms for Blade Cutting No. 16 Cord Protected by an AFCI.
Available Short Circuit Current = 100A

Courtesy: Cutler - Hammer

143

Figure 40 is a schematic diagram of how the Cutler- Hammer single pole, 120v Branch/Feeder AFCI operates.

Figure 40

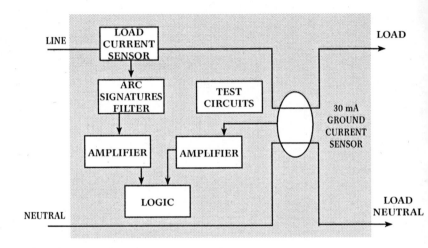

Single Pole Branch/Feeder Arc Fault Circuit Interrupter

Courtesy: Cutler - Hammer

The load current sensor sends information to the arc signature filter, which decides whether the arc signature should be sent on for processing based on the magnitude of the arcing currents. Normal, non- arc related currents are filtered out. If the information is sent on, it is amplified, and sent to the logic circuitry. A decision is made as to whether the arc is a "good arc" or a "bad arc". It distinguishes between the wave forms of normal circuit transients including those associated with incandescent lamp burn- out, and the wave forms of dangerous arcing events. If it is determined to be a "bad arc", the breaker is

tripped by an output from the logic unit. This circuitry is used
to protect against parallel arc faults.

Note that the device has a toroid, or current transformer which
provides 30 milliampere GFP – like protection. This is how the
device protects against series arcs in NM-B cable, and also protects against arcing faults to ground. In the case of NM-B cable,
the device detects leakage currents to the equipment grounding
conductor associated with the series arcing condition.

There is a test circuit which, when the test button is depressed,
introduces a simulated "bad arc" into the system, tripping the AFCI.

Figure 41 shows the AFCI combined with the GFCI function.
The only significant difference is the addition of a second current transformer for the double grounded neutral protection and
the lowering of the residual current trip level to 6ma- Must trip
at 6ma, must not trip below 4ma. In addition to the AFCI test
button, this device would have an additional test circuit with the
same test button setting as regular GFCIs, and would be dual-
listed as an AFCI and a GFCI.

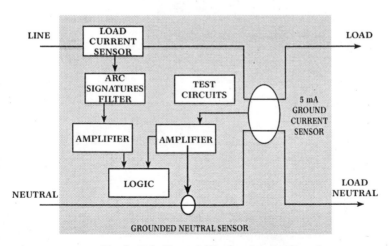

Single Pole Branch/Feeder Arc Fault and
Ground Fault Circuit Interrupter

Figure 41

Courtesy: Cutler - Hammer

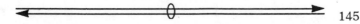

Figure 42 shows a two- pole, combination AFCI and GFCI. It could easily become a two- pole AFCI and GFP – like breaker by eliminating the double - grounded neutral sensor and changing the residual current trip level to 30ma or some other value.

Notice that there are test circuits for each pole. The breaker has a common trip.

Figure 42

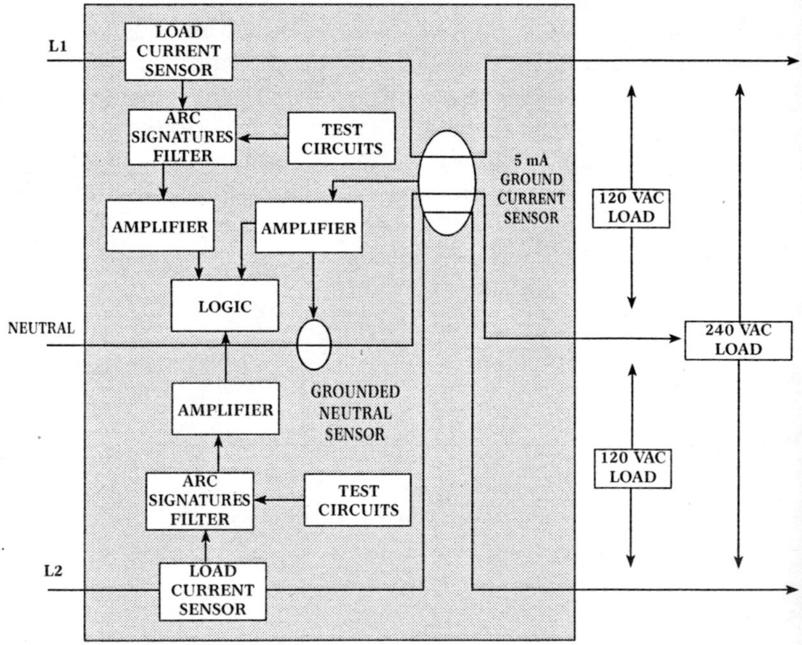

Two Pole Branch/Feeder Arc Fault and
Ground Fault Circuit Interrupter

Courtesy: Cutler - Hammer

It is noted in a summary of the UL 1699 tests in Figure 36 that there are extensive tests to determine that the device will not nuisance trip under various load conditions such as inrush currents, normal operation arcing, non-sinusoidal waveforms, cross talk, multiple loads, and lamp burnout.

The device is also subjected to various conditions which might inhibit its ability to trip, such as "masking", or signal blocking by other electrical or electronic devices, and EMI, or electrical - magnetic interference filtering.

Figure 43 shows various Cutler – Hammer miniature circuit breakers with the additional AFCI function. Some of these circuit breakers contain both UL listed AFCI and GFCI functions.

Figure 43

Various AFCI plus GFCI Circuit Breaker Devices

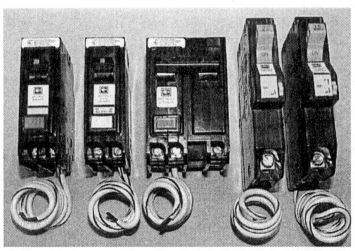

Courtesy: Cutler - Hammer

It's early in the evolutionary cycle of arc – fault circuit – interrupters, but AFCIs appear to have a promising future, in adding a significant new element to the safety of our electrical systems.

Chapter 17
The Future

Most of us can remember a handful of teachers during our years through the educational process who really made a lasting impact on us. One who made such an impact on me forced my mind to have certain bits of poetry engraved on it. I think of her every time one of the poems pops into my conscience. Here's one that applies to my attempt to predict the future:

> "For I dipped into the future, far as human eye could see.
> Saw a vision of the World, and all the wonders that will be."

Another gem is probably more appropriate to any attempt to predict the future:

> "A little knowledge is a dangerous thing.
> Drink deeply, or taste not the Pyhrean Spring.
> There, shallow drafts intoxicate the brain;
> And drinking deeply, sobers us, again."

Oh, well, here goes. I'm sure we'll all have a good laugh about some of what I predict, when we review it in a few years.

I see a continued progressive growth for residual current devices of all types and the introduction and growth of new types. Eventually, I believe RCD's will be used almost universally at services, feeders, branch circuits, and receptacles. Most appliances will have some type of electric shock, and, in some cases, fire protection devices built into them.

Previously, I mentioned that for quite a few years, I had access to a national newspaper clipping service for all electric shock and electrical fire newspaper articles. In analyzing this information for a specific two year period, I noticed that electric shock accidents occurred just about everywhere around the home except the bedroom. I commented in my report that "the accidents which occur in the bedroom appear to be other than electrical in nature."

Later information has proven that this observation, although applicable to the information reviewed, was too optimistic. Although rare, electrocutions do occur also in the bedroom, and GFCI protection has merit there, as well as in other areas of the home.

The Long Range Effects of the AFCI

Within Buildings :

During the next decade, it is likely that the AFCI will be used extensively to protect most of the 15A and 20A branch circuits in dwellings, nursing homes, senior care homes, schools, motels, and hotels. It is noteworthy that smoke alarms do a great job of alerting people to the existence of a fire, but they don't prevent a fire from starting. The AFCI can PREVENT a fire from occurring.

In the long – long range future, decades from now, it is likely that the AFCI principle will be refined, and the ratings of AFCIs will be extended. AFCIs will be applied throughout the electrical systems of residential, commercial, and industrial buildings.

It is likely that the AFCI will be combined with the GFCI, GFP, circuit breaker, receptacle, plug, cord connector, appliances, etc. to offer " circuit protection" throughout all necessary locations of all buildings.

The name of the game will change from "overcurrent protection", "GFCI protection", "GFP protection", and "AFCI protection" to "circuit protection". Circuit protection will incorporate various combinations of the various protection types, as required for the particular circuit need. Of course, the way our NE Code process works, it will take many Code cycles to arrive at such an advanced state of development.

At Services :

One of the remaining weak points in our building electrical systems is the service to the building. Would you believe that there are NO NEC requirements for providing overcurrent protection to any of the service entrance conductors of the millions of buildings in the USA, and that the vast majority of buildings have no such protection?

Since the NEC was first written, until the 1987 NEC, Section 230-90 stated that each ungrounded service- entrance conductor "shall have overcurrent protection." Section 230-90(a) stated that such protection shall be provided by an overcurrent device in series with each ungrounded service conductor having a rating or setting not higher than the allowable ampacity of the conductor. Of course, these overcurrent devices were on the LOAD side of the service- entrance conductors.

When I finally became aware of the fact that this was really only OVERLOAD protection and didn't provide protection against short circuits or high level ground faults on the LINE side of the overcurrent device, I attempted to correct the illusion. I rode in on an innocuous proposal, No. 4-167 in the 1986 TCR, and submitted Comment No. 4-85, on Page 74 of the 1986 TCD. The result was that the word "overcurrent" was changed to "overload", which is correct. The point here is that we still don't have overcurrent protection. Not only should we have overcurrent protection, but, in the future, we should have AFCI protection, as well. Granted that the residential fire data quoted in Chapter 16 doesn't indicate a pressing need, but there ARE fires on the line side of building services, particularly commercial and industrial buildings. The CPSC study did not cover commercial and industrial buildings.

Presently, we clear the faults by burning off the conductors. In essence, the conductors become fuses. It would seem that in the future, a device providing both AFCI and overcurrent protection will be devised to address this problem.

It is interesting to observe that the combination AFCI which has been UL listed is reported to be capable of detecting a series arc condition on the LINE side of the device, and disconnecting the loads protected by the device. Such a series arc fault could be from a broken conductor, a corroded terminal connection at a meter, on a switch, fuse, or circuit breaker, etc. The horrendous experiences the electrical industry went through with aluminum conductors, particularly when they spread to 15A and 20A branch circuits, were mostly series arc fault and glowing terminal problems.

The combination rated AFCI would prevent the series arc fault from progressing to a parallel arc fault or an arc fault to ground by disconnecting the load, and thus, disconnecting the series arc current. Of course, once the arc fault progressed to a parallel arc fault or an arc fault to ground, it would require a protective device on the LINE side of the fault to disconnect the circuit. This is where an AFCI- type device at the service point would have an application.

The present AFCI standard, UL 1699, covers devices rated only 15 and 20A, 120V. As the technology advances and the higher ratings become available, we could have devices at the services of buildings which are easily inserted into the service point of the buildings to give the needed protection. When a fault occurred, the power would be disconnected. After the condition causing the fault was corrected, the device could be reset by a remote control device like our TV or garage door controller.

It would be necessary to be sure that arcing faults and other faults downstream on the system did not unnecessarily disconnect power to the whole building, but this type of problem is solved now be judicious sizing and setting of the devices at the feeder and branch circuit levels to activate the device closest to the fault.

I'm looking forward to seeing these devices required at services in the 2050 issue of the National Electrical Code.

Electric Vehicles

One of the potentially large areas for the growth of RCDs, particularly GFCIs of new designs, will be contributing to the safety of electric vehicles.

I remember way back in the 1950's when I was with an electric utility, I was invited to attend the annual meeting of the EEI, the Edison Electric Institute, in Atlantic City. The utilities had commited to buy 1000 electrified Renaults, and we drove one up and down the Boardwalk.

About all that experiment proved was that you can't take a regular car, fill it with lead-acid batteries, and have a viable product. A breakthrough was needed in battery technology and systems. Since then, there have been many changes in all phases of electric vehicle design, and the industry, prodded by California anti-polution edicts, appears to have viable vehicles.

The NE Code adapts readily to the needs of new technology, and the 1996 NE Code has a new Article 625, Electric Vehicle Charging System Equipment. Section 625-22 of the 1996 Code requires that all electric vehicle supply equipment have GFCI protection. Where cord- and plug-connected electric vehicle supply equipment is used, the GFCI protection is to be in the plug or within 1 ft of the plug.

The developments on electric vehicle chargers are focusing around three basic charging levels:

Slow Charge-15 A, 120 v AC from present receptacles. Use present GFCIs.

Normal Charge-30 A, 240 v AC derived from a 240/120 v AC supply with: 120 v to ground, or 208 v AC derived from a 208Y/120 v 3-phase supply with 120 v to ground. Use present GFCIs.

Fast Charge-75 to 200 kvA-Not yet defined, but likely available at electric "filling stations." Could be at 480Y/277 v AC 3 phase, or 250 v DC. New type of protection needed. A new standard, UL 2231, addresses this need.

In order to match the safety level of present GFCIs, any 480v device would have to interrupt the current in 2 milliseconds-much faster than present technology has achieved. For DC, no known GFCI technology exists.

Present GFCI parameters are based on a 120 v AC nominal

voltage plus a 10% tolerance, which gives us 132 v. The lowest "normal" body impedance is taken as 500 ohms. The corresponding "worst case" body fault current comes out to: I = E/Z = 132/500 = 0.264 A, or 264 mA.

For voltages of 150 v or less to ground, the tripping time of a GFCI must be less than 25 milliseconds, at 264 mA. For the low end of the trip current range, at 6 mA the GFCI has 5.6 seconds to trip. As indicated previously, at 6 mA, Class A GFCIs actually trip much faster than this, even less than 1/10th of a second, as shown in Fig. 14.

As the technical parameters are resolved for the Fast Charge units, solutions will be found to meet them. Progress is wedded to the future of the electric vehicle. If the electric vehicle is truly here to stay, the personal protection products will be available.

Intelligent Buildings

This subject is split among residential, commercial, and industrial buildings. All have the potential to be made much more "intelligent" from many standpoints including energy conservation, appliance controls, and electric shock and electrical fire safety.

Residential Buildings

Varying degrees of sophistication in home wiring have been around for a long time. I used to work for a company which had an elaborate low-voltage remote control wiring system. I installed it in my home and added prototype devices to it. The system was a big success with our top company executives, because we provided system design, installation, and maintenance for their home systems by highly-capable engineers, free of charge. In the real world it was only marginally successful. A heavier duty commercial building variation of the system had much more success.

I mention this because even with a basically straightforward, nonelectronic system there can be a big engineering input required in design, installation, and trouble-shooting. The most ambitious residential "intelligent" building project was the "Smart House" project sponsored by the National Association of Home Builders. A number of model homes have been built, and other developments are underway. I wish them well. Two major subjects I see that they must overcome are the problem of convincing the appliance industry to manufacture both "smart" appliances and the present appliances, plus the mobility of the American people, and the resulting mix of "smart" and "dumb" appliances. Also, the electronic trouble-shooting problems could become astronomical.

I remember many years ago, we were attempting to introduce a line of new garage door operators to the market. As usual, I had installed a prototype in my garage. The landing pattern for the local airport was right over my house. Regularly, the garage door would go up or down unexpectedly from the radio transmissions from the airplanes. The problem was resolved, and such problems with modern door operators are very rare. The point is that when dealing with electronics, funny things happen, sometimes requiring field investigations by skilled people to resolve. I'm convinced that there will be some technical home wiring advances as a result of the "Smart House" project, but only time will tell us the degree of acceptance.

The 1987 NE Code responded to the special wiring needs of the "Smart House" project, and added Article 780, Closed-Loop and Programmed Power Distribution, specifically for the project. The Article has been revised in each NE Code since, to track the changes being incorporated into the system.

On a less ambitious scale, residential loadcenters and panelboards have been introduced which incorporate certain "intelligent loadcenter" features, usually including surge protection, shunt trip breakers, and GFP breakers.

A major change in the 1996 NEC concerning how electric ranges and dryers can be grounded could open up many opportunities for applying RCDs at the service entrance, protecting the entire building or large segments of it. The previous NE Codes from WW II on allowed the grounded-circuit conductor, or neutral, to serve also as the equipment grounding conductor.

A check of the records shows that this allowance was made during the war only " for the duration", but there was no definition of ' duration". This became the more common wiring method. This in essence created a second ground on the conductor, and would cause any GFCI with the required double-grounded neutral protection to trip, making it impossible to apply conventional GFCIs at such locations.

The 1996 NEC in Section 250-60 eliminates the use of 3-wire electric range and clothes dryer circuits for new construction, requiring a separate equipment grounding conductor like all other circuitry. GFCIs can now be used with these appliances.

Heating Elements

Sheathed heating elements, such as those used as surface units on electric ranges and many other applications, often use a conductor surrounded by a compressed layer of magnesium oxide, enclosed in a metal sheath. The total element is hermetically sealed except where the conductor exits the unit to connect to the electrical supply. It has been stated that the seal at the wire exit is not hermetically sealed to allow the element to "breathe." Introduction of new oxygen from the air improves the likelyhood that the magnesium oxide will remain chemically stable, and not revert to magnesium and oxygen. If it does revert, the magnesium is a conducting metal and the unit fails. When a heating element on a range top fails, it can fail catastrophically, with a persistent arcing ground fault that can cause a fire very difficult to extinguish. Other sheathed heating elements in other appliances can experience similar failures.

One of the problems with trying to protect against such a failure by a GFCI is the inherently high "standing leakage" of such heating elements. Because they "breathe" they can take in moist air when they are cold, causing electrical leakage as high as 100 mA when they are first turned on. After they heat up, the moisture is driven off, and the leakage current goes down to a low value, usually below 1 mA.

I predict that RCD type devices with extra–intelligence will evolve which can be programmed to recognize the "normal" profile of devices and components with unusual but predictable electrical leakage. Such an RCD could be designed to trip when leakage current exceeded the predicted profile. It could also be programmed to give a visual or audible warning at some level of leakage current, and to trip at a more dangerous level of current. This could be like the leakage monitoring receptacle or LMR, discussed in Chapter 15, or possibly an even more sophisticated device.

I attempted once to interest the range industry in such a device, but to no avail. The public and the appliance industry in response to the public, appeared more interested in features like ice water through the door of a refrigerator than in preventing catastrophic failures of heating elements.

Commercial Buildings

The "intelligent building" has progressed among commercial buildings, particularly concerning energy conservation. As the electronic control capabilities of commercial buildings continue, more electric shock and electrical fire protection capabilities are bound to appear. Requirements for GFCI and GFP protection are spreading into commercial occupancies, and this should continue.

Industrial Buildings

Industrial buildings are most likely to have skilled personnel on site to install and maintain electronic devices and systems. Where the application of a device or system can be demonstrated as saving cost, increasing productivity, or improving safety, it is most likely to be accepted. The pressures of OSHA and our litigious society add additional incentives.

In conclusion, I see the application of electronics in the form of the GFCI, GFP, AFCI, ALCI, ELCI, IDCI, LMR, and other devices and systems yet unnamed, as continuing the revolution which is underway in electric shock and electrical fire protection.

Since the introduction of GFCIs, the average annual death rate in the USA from electrocutions by utilization voltages has been cut in half, from somewhat over 600 per year to about 300. GFP has contributed to a decrease in electrical fires, with resultant savings in life and property damage.

The intriguing possibility exists for extending the AFCI technology to electrical systems other than building electrical systems. A U.S. Government study is underway to evaluate the application of AFCIs to aircraft electrical systems.

The results of this study could lead to wider use of AFCI devices, particularly in aircraft.

In the future, dangerous Undercurrents will be addressed with the same degree of concern which we have given to Overcurrents, and a drop in the annual death rate from electric shock and electrical fires, as well as property damage from electrical fires, should continue.

Chapter 18

Definitions of Terms

Adapter – A wiring device with plug blades and one or more receptacles that converts from one wiring device configuration to another.

AFCI – Arc-Fault Circuit-Interrupter. A device intended to provide protection from the effects of arc faults by recognizing characteristics unique to arcing and by functioning to de-energize the circuit when an arc fault is detected.

ALCI – Appliance leakage circuit interrupter. A residual current device which provides personnel electric shock protection equivalent to a Class A GFCI, but lacks the double–grounded neutral tripping feature and is suitable for use only as a component of a listed product.

Arcing – A luminous discharge of electricity across an insulating medium, usually accompanied by the partial volatilization of the electrodes.

Arcing Fault – An unintentional arcing condition in a circuit.

Bonding – The permanent joining of metallic parts to form an electrically conductive path that will assure electrical continuity and the capacity to conduct safely any current likely to be imposed.

Branch Circuit – The circuit conductors between the final overcurrent device protecting the circuit and the outlet(s).

Branch/Feeder Arc-Fault Circuit-Interrupter – A device intended to be installed at the origin of a branch circuit or feeder, such as at a panelboard. It is intended to provide protection of the branch circuit wiring, feeder wiring, or both, against unwanted effects of arcing. The device also provides limited protection to branch circuit extension wiring. It may be a circuit-breaker type device or a device in its own enclosure mounted at or near a panelboard.

Carbonized Path – A conductive carbon path formed through or over the surface of a normally insulating material.

Circuit Breaker – A device designed to open and close a circuit by nonautomatic means and to open the circuit automatically on a predetermined overcurrent without damage to itself when properly applied within its rating.

Combination Arc-Fault Circuit-Interrupter – An AFCI which complies with the requirements for both branch/feeder and outlet circuit AFCIs. It is intended to protect downstream branch circuit wiring and cord sets and power-supply cords.

Cord Arc-Fault Circuit-Interrupter – A plug-in device intended to be connected to a receptacle outlet. It is intended to provide protection to the power-supply cord connected to it against the unwanted effects of arcing. The cord may be integral to the device. The device has no additional outlets.

Cord Connector – A wiring device designed to be attached to a cord, with one or more integral receptacles.

ELCI – Equipment leakage circuit interrupter. A residual current device which is similar to an ALCI, except that its minimum trip level is set higher. Whereas the ALCI must trip at 6 mA, and must not trip below 4 mA in order to provide the personnel protection of a Class A GFCI, the ELCI can have a trip level of 20 mA, 30 mA, or some other value to meet the needs of the equipment it is protecting.

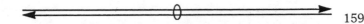

Feeder – All circuit conductors between the service equipment or the source of a separately derived system and the final branch circuit overcurrent device.

Fuse – A device with one or more fuseable elements which provides overcurrent protection when properly applied within its rating.

GFCI or GFI – Ground fault circuit interrupter. A device intended for the protection of personnel that functions to de–energize a circuit or portion thereof within an established period of time when current to ground exceeds some predetermined value that is less than that required to operate the overcurrent protective device of the supply circuit.

Although this is the official definition, it should be kept in mind that the GFCI really responds to "RESIDUAL" currents, of which ground faults are the most likely cause.

GFP or GFPE – Ground fault protection of equipment. A system intended to provide protection of equipment from damaging line–to–ground fault currents by operating to cause a disconnecting means to open all ungrounded conductors of the faulted circuit. The protection is provided at current levels less than those required to protect conductors from damage through the operation of a supply circuit overcurrent device.

Ground Fault – An unintentional electrical connection between an energized conductor and ground.

IDCI – Immersion detection circuit interrupter. A device designed to disconnect the electrical supply to an appliance at the power supply plug of the appliance when the appliance becomes immersed in water.

Lighting Outlet – An outlet intended for the direct connection of a lampholder, a lighting fixture, or a pendant cord terminating in a lampholder.

LMR – Leakage monitoring receptacle. A device which uses the residual current principle and which detects one or more predetermined levels of leakage current. The LMR announces the existence of the leakage current by visual and/or audible means. It may be combined with the ELCI function, and trip at some higher level of leakage current.

Operation Inhibition – Denotes the concealment of an arcing fault by the normal operation of certain circuit components.

Outlet – A point on a wiring system at which current is taken to supply utilization equipment.

Outlet Circuit Arc-Fault Circuit-Interrupter – A device intended to be installed at a branch circuit outlet, such as at an outlet box. It is intended to provide protection of cord sets and power-supply cords connected to it (when provided with receptacle outlets) against the unwanted effects of arcing. This device may provide feed-through protection of the cord sets and power-supply cords connected to downstream receptacles.

Overcurrent Protection – The sensing and interrupting of overload, short circuit, and high level ground fault currents.

Overload – Operation of equipment in excess of normal full load rating, or of a conductor in excess of rated ampacity, that when persistent for a sufficient length of time causes damage or overheating.

Personnel Protection System – A system of personnel protection devices and constructional features that when used together provide protection against electric shock of personnel.
The definition became necessary for electric vehicle charging systems, where the higher voltages necessary to accommo-

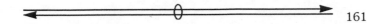

date fast charging make it impractical to meet the normal Class A GFCI parameters. The definition provides greater latitude in designing a safe installation.

Plug, Attachment Plug, Plug Cap, or Cap – A device that by insertion in a receptacle, establishes connection between the conductors of the attached flexible cord and the conductors connected permanently to the receptacle.

Portable Arc-Fault Circuit-Interrupter – A plug-in device intended to be connected to a receptacle outlet and provided with one or more outlets. It is intended to provide protection to connected cord sets and power-supply cords against the unwanted effects of arcing.

RCD – Residual Current Device. A device that constantly monitors the current flowing in a circuit and opens all ungrounded conductors when a residual current reaches a predetermined level.

Receptacle – A contact device installed at the outlet for the connection of a single attachment plug.

Receptacle Outlet – An outlet where one or more receptacles are installed.

Short Circuit – An unintentional connection between two or more circuit conductors.

Unwanted Trip – AFCI - A tripping function in response to a condition that is not an arcing fault but a condition that occurs as part of the normal or anticipated operation of circuit components.

Unwanted Trip - GFCI and GFP – A tripping function in response to a condition that is not a residual current condition that the device is intended to respond to, as defined by the UL listing process.

Acknowledgements

A special word of thanks is extended by the author to the following :

Prof. Theodore Bernstein - For contributing historical insights concerning the Edison versus Westinghouse disputes and the creation and early usage of the electric chair.

Prof. Dr. Gottfried Biegelmeier - For insight into the effects of electric current on the human body.

Dr. Norman Boas, MD - For reviewing a draft of portions of the text and for contributing to the text's medical information.

Steve Campolo - For information on recent GFCI developments.

Canadian Standards Association - CSA - For the World Electrical System data in the Appendix.

Bob Clarey - For help in understanding and sharing information on the AFCI.

Viv Cohen - For insight into the history and usage of RCDs in South Africa.

Phil Cox - For a comprehensive review of a draft of the text.

General Electric Co. - For use of many tid- bits of information gained during the author's career with GE.

Bernard Gershen - For sharing information on GFCIs and similar devices.

International Electrotechnical Commission - (IEC) - Parts of IEC 479 are reproduced by kind permission of the International Electrotechnical Commission (IEC). Copies of the complete standard may be obtained from IEC, P.O. Box 131, 1211 Geneva 20, Switzerland, Phone : + 41 22 919 0228, Fax : + 41 22 919 0300 or, in the United States, from the American National Standards

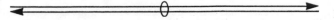

Institute, Sales Department, 11 West 42nd Street, New York, NY-10036, Phone : + 1 212 642 4900, Fax : + 1 212 302 1286.

Clive Kimblin - For help in understanding and sharing information on the AFCI.

Bill King - For assistance with CPSC fire and electric shock data.

Ray Legatti - For historical information on the early development of GFCI electronic modules.

Leviton Mfg. Co., Inc. - For their support and encouragement in the creation of the book.

National Electrical Manufacturers Association - NEMA - For permission to use the information from NEMA publications, and for the cooperation received from many of the NEMA - member companies.

National Fire Protection Association - NFPA - For permission to use the many references in the text to the National Electrical Code and NFPA.

Bill Nestor - For historical insight into the creation of the first GFCIs.

George Ockuly - For historical and current information on fuse technology.

James N. Pearse - For a comprehensive critique of a draft of the text.

Saul Rosenbaum - For anecdotal contributions and technical support.

Ned Schiff - For information on the Fire Shield device.

Charles B. Schram - For reviewing a draft of the text and for suggesting improvements.

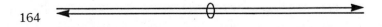

Walter Skuggevig - For reviewing a draft of portions of the text and contributing information on the effects of electric current on people.

John Wafer - For information on the creation of molded case circuit breakers and on arc fault circuit interrupters.

Pat Ward - For contributing information on the workings of European electro- mechanical RCD's.

Jack Wells - For insight into the early development and industry acceptance of GFCIs.

Bibliography

CSA Standards - Canadian Standards Association, 178 Rexdal-Blvd., Rexdale (Toronto) , Ontario, Canada M9W1R3 :

>   Canadian Electrical Code - Part 1
>   CSA Standard C 22.1

>   Ground Fault Circuit Interrupters
>   CAN/CSA C22.2 No. 144-M91
>   A National Standard of Canada

IEC Standards - International Electrotechnical Commission, IEC Central Office - 3, rue de Varembe, 1211 Geneva 20, Switzerland:

>   Effects of Current Passing Through the Human Body-
>   IEC Publication 479

NEMA Publications - National Electrical Manufacturers Association 1300 N. 17th St., Suite 1847, Rosslyn, VA 22209:

>   Application Guide for Ground Fault Circuit Interrupters
>   Standards Publication No. 280

>   Molded Case Circuit Breakers and Molded Case Switches
>   Standards Publication No. AB 1

>   Molded Case Circuit Breakers and Their Application
>   Standards Publication No. AB 3

>   Proceedure for Evaluating Ground Fault Circuit Interrupters
>   for Response to Conducted Radio Frequency Energy
>   Standards Publication No. PP 1

NFPA Standards - National Fire Protection Association, 1 Batterymarch Park, PO Box 9101, Quincy, MA 02269-9101 :

>   National Electrical Code

UL Standards - Underwriters' Laboratories, Inc., 333 Pfingsten Rd., Northbrook, IL 60062-2096 :

Standard for Safety - Personnel Grooming Appliances - UL 859

Standard for Safety - Ground-Fault Circuit-Interrupters - UL 943

Standard for Safety - Immersion-Detection Circuit-Interrupters - UL 1664

Standard for Safety - Arc Fault Circuit-Interrupters - UL 1699

Standard for Safety - Personnal Protection Systems for Electric Vehicle Supply Circuits - UL 2231

World Electrical Supply Systems
Data — Courtesy of Canadian Standards Association

COUNTRY	FREQUENCY AND TOLERANCE Hz & %	HOUSEHOLD VOLTAGE V	COMMERCIAL VOLTAGE V	INDUSTRIAL VOLTAGE V	VOLTAGE TOLLERANCE V
AFGHANSTAN	50	380/220 (A) 220 (L)	380/220 (A)	380/220 (A) (3)	(9)
ALGERIA	50 ± 1.5	220 (L) (1) 220/127 (E)	380/220 (A) 220/127 (A)	10kV 5.5kV 380/220 (A)	± 5 to 10
ANGOLA	50	220 (L) (1)	380/220 (A)	380/220 (A) (3)	(9)
ANTIGUA	60	230 (L) (1)	400/230 (A)	400/230 (A)	(9)
ARGENTINA	50 ± 1.0	225 (L) (1) 220 (L) (1)	390/225 (A) 380/220 (A) 220 (L)	13.2kV 6.88kV 390/225 (A) 380/220 (A)	± 10
AUSTRALIA	50 ± 0.1	415/240 (A) (E) 240 (L)	415/240 (A) 440/250 (A) 440 (N) (6)	22kV 11kV 6.6kV 415/240 (A) 440/250 (A)	± 6
AUSTRIA	50 ± 0.1	380/220 (A) (B) 220 (L)	380/220 (A) (B) 220 (L)	20kV 10kV 5kV 380/220 (A)	±5
BAHAMAS	60	240/120 (G) 120 (L)	240/120 (G) 120 (L)	415/240 (A) (3) 208/120 (A)	(9)
BAHRAIN	50 & 60	400/230 (A) 230 (L) 110 (L)	400/230 (A) 380/220 (A) 230 (L) 220/110 (K)	11kV 400/230 (A) 380/220 (A)	± 6
BANGLADESH	50 ± 4	400/230 (A) 230 (L)	11kV 400/230 (A)	11kV 400/230 (A)	± 5
BARBADOS	50 ± 0.4	230/115 (G) (K) 200/115 (A) (E)	230/115 (G) (K) 200/115 (A) (E)	11kV 3.3kV 230/115 (G) 200/115 (A)	± 6
BELGIUM	50 ± 3	380/220 (A) 220/127 (A) 220 (F)	380/22/ (A) 220/127 (A) 220 (F)	15kV 6kV 380/220 (A) 220/127 (A) 220 (F)	± 5 (day) ± 10 (night)
BELIZE	60	220/110 (K)	220/110 (K)	440/220 (A) (3)	(9)
BERMUDA	60 ± 0.1	240/120 (K) 208/120 (A)	240/120 (K) 208/120 (A)	4.16/2.4kV 208/120 (A) 240/120 (K)	± 5
BOLIVIA	50 ± 1	230/115 (H)	230/115 (H)	230/115 (H) (3)	± 5
BOTSWANA	50	220 (L) (1)	380/220 (A)	380/220 (A) (3)	(9)
BRAZIL	60	220 (L) (1) 127 (L) (1)	380/220 (A) 220/127 (A)	13.8 kV 11.2kV 380/220 (A) 220/127 (A)	(9)
BULGARIA	50 ± 0.1	380/220 (A) 220 (L)	380/220 (A) 220 (L)	20kV 15kV 380/220 (A)	± 5
BURMA	50	230 (L) (1)	400/230 (A) 230 (L)	11kV 6.6kV 400/230 (A)	(9)
CAMBODIA	50	208/120 (A) 120 (L)	380/220 (A) 208/120 (A)	380/220 (A) (3) 208/120 (A)	(9)

World Electrical Supply Systems
Data — Courtesy of Canadian Standards Association

COUNTRY	FREQUENCY AND TOLERANCE Hz & %	HOUSEHOLD VOLTAGE V	COMMERCIAL VOLTAGE V	INDUSTRIAL VOLTAGE V	VOLTAGE TOLLERANCE V
CAMEROON (FR)	50 ± 2	200 (L) (1)	380/220 (A)	15kV 380/220 (A)	± 5
CANADA	60 ± 0.02	240/120 (K)	600/347 (A) 480 (F) 240 (F) 240/120 (K) 208/120 (A) (G) 416/240 (A) (D)	12.5/7.2kV 600/347 (A) 208/120 (A) 600 (F) 480 (F) 240 (F)	+ 4 -8.3
CAYMAN ISLANDS	60 ± 0.01	240/120 (K)	240/120 (K) (G)	480/240 (G) 480/227 (A) 240/120 (G) 208/120 (A)	± 10
CHAD	50	220 (L) (1)	220 (L) (1)	380/220 (A) (3)	(9)
CHILI	50	220 (L) (1)	380/220 (A) (1)	380/220 (A) (3)	(9)
CHINA (PR)	50	220 (L)	380/220 (A)	380/220 (A)	± 10
COLOMBIA	60 ± 1	240/120 (G) 120(L)	240/120 (G) 120(L)	13.2kV 240/120 (G)	± 10
COSTA RICA	60	120 (L) (1)	240/120 (K)	240/120 (G) (3)	(9)
CYPRUS	50 ± 2.5	240 (L) (1)	240 (L) (1)	11kV 415/240 (A)	± 6
CZECHOSLOVAKIA	50 ± 0.1	380/220 (A) 220 (L)	380/220 (A) 220 (L)	22kV 15kV 6kV 3kV 380/220 (A)	± 10
DENMARK	50 ± 0.4	380/220 (A) 220 (L)	380/220 (A) 220 (L)	30kV 10kV 380/220 (A)	± 10
DAHOMEY	50 ± 1	380/220 (A) 220 (L)	380/220 (A) 220 (L)	15kV 380/220 (A)	± 10
OOMINICAN REPUBLIC	60	110 (L) (1)	220/110 (K) (1) 110(L)	220/110 (G) (3)	(9)
ECUADOR	60	127 (L) (1) 120 (L) (1) 110 (L)	240/120 (K) 208/120 (A) 220/127 (A) 220/110 (K)	240/120 (K) 208/120 (A) 220/127 (A) 220/110 (K)	(9)
EGYPT (AR)	50 ± 1	380/220 (A) 220 (L)	380/220 (A) 220 (L)	11kV 6.6kV 380/220 (A)	± 10
EL SALVADOR	60 ± 1	240/120 (K)	240/120 (K) (G)	14.4 kV 2.4kV 240/210 (G)	± 5
ETHIOPIA	50	220 (L) (1)	380/220 (A)	380/220 (A) (3)	(9)
FALKLAND ISLANDS	50 ± 3	230 (L) (1)	415/230 (A)	415/230 (A) (3)	± 2.5
FIJI ISLANDS	50 ± 1	415/240 (A) 240 (L)	415/240 (A) 240 (L)	11kV 415/240 (A)	(9)
FINLAND	50 ± 0.1	220 (L) (1)	380/220 ()	660/380 (A) 500 (B) 380/220 (A) (D)	± 10
FRANCE	50 ± 1	380/220 (A) 220 (L) 220/127 (E) 127 (L)	380/220 (A) 380/220 (D) 380 (B)	20kV 15kV 380 (B) 380/220 (A) (D)	± 10
GAMBIA	50	230 (A) (1)	230 (A) (1)	400/230 (A) (3)	± 5 (1)

World Electrical Supply Systems
Data — Courtesy of Canadian Standards Association

COUNTRY	FREQUENCY AND TOLERANCE Hz & %	HOUSEHOLD VOLTAGE V	COMMERCIAL VOLTAGE V	INDUSTRIAL VOLTAGE V	VOLTAGE TOLLERANCE V
GERMANY (FR)	50 ± 0.3	380/220 (A) 220 (L)	380/220 (A) 220 (L)	20kV 10kV 6kV 380/220 (A)	± 10
GERMANY (DDR)	50 ± 0.3	380/220 (A) 220 (L) 220/127 (A) 127 (L)	380/220 (A) 220 (L)	10kV 6kV 660/380 (A) 380/220 (A)	± 5
GHANA	50 ± 5	250 (L) (1)	250 (L) (1)	440/250 (A) (3)	± 10
GIBRALTAR	50 ± 1	415/240 (A)	415/240 (A)	415/240 (A) (3)	± 6
GREECE	50 ± 1	220 (L) (1)	6 6kV 380/220 (A)	22kV 20kV 15kV 6.6kV 380/220 (A)	± 5
GRENADA	50	230 (L) (1)	400/230 (A)	400/230 (A) (3)	(9)
GUADALOUPE	50 & 60	220 (L) (1)	380/220 (A)	20kV 380/220 (A)	(9)
GUAM (Mariana Islands)	60 ± 1 -0.08	240/120 (K) 208/120 (A) 240 (L) 120 (L)	240/120 (K) 208/120 (A)	13.8kV 4.0kV 480/277 (A) 480 (D) 240/120 (H) 208/120 (A)	+ 8 10
GUATEMALA	60 ± 1.7	240/120 (K)	240/120 (K)	13.8kV 240/120 (G)	± 10
HAITI	60	230 (L) (1) 220 (L) (1) 115 (L)	380/220 (A) 230/115 (K) 220 (L)	380/220 (A) 230/115 (G)	(9)
HONDURAS	60	110 (L)	220/110 (K) 110 (L)	220/110 K) (3)	(9)
HONG KONG (and Kowloon)	50 ± 2	346/200 (A) 200 (L)	11kV 346/200 (A) 200 (L)	11kV 346/200 (A)	± 6
HUNGARY	50 ± 2	380/220 (A) 220 (L)	380/220 (A) 220 (L)	20kV 10kV 380/220 (A)	+ 5 -10
ICELAND	50 ± 0.1	380/220 (A) 220 (L)	380/220 (A) 220 (L)	380/220 (A) (3)	(9)
INDIA (4)					
Bombay	50 ± 1	440/250 (A) 230 (L)	440/250 (A) 230 (L)	11kV 440/250 (A)	± 4
New Delhi	50 ± 3	400/230 (A) 230 (L)	400/230 (A) 230 (L)	11kV 400/230 (A)	± 6
Ramakrishnapuram (2)	50 ± 3 25 d.c.	400/230 (A) 230 (L) 460/230 (P)	400/230 (A) 230 (L) 460/230 (P)	22kV & 11kV (9) (9)	± 6
INDONESIA	50 + 1	220/127 (A)	380/220 (A) 220/127 (A)	380/220 (A) (3)	± 5
IRAN	50 ± 5	220 (L) (1)	380/220 (A)	20kV 11kV 400/231 (A) 380/220 (A)	± 15
ᴵᴿᴬᴏ	50	220 (L) (1)	380/220 (A)	11kV 380/220 (A)	(9)

World Electrical Supply Systems
Data — Courtesy of Canadian Standards Association

COUNTRY	FREQUENCY AND TOLERANCE Hz & %	HOUSEHOLD VOLTAGE V	COMMERCIAL VOLTAGE V	INDUSTRIAL VOLTAGE V	VOLTAGE TOLLERANCE V
IRELAND NORTHERN	50 ± 0.4	230 (L) 220 (L) (1)	400/230 (A) 380/220 (A)	400.230 (A) (3) 380/220 (A)	± 6
IRELAND REPUBLIC OF	50	220 (L) (1)	380/220 (A)	10kV 380/220 (A)	(9)
ISRAEL	50 ± 0.2	400/230 (A) 230 (L)	400/230 (A) 230 (L)	22kV 12.6kV 6.3kV 400/230 (A)	± 6
ITALY	50 ± 0.4	380/220 (A) 220/127 (E) 220 (L)	380/220 (A) 220/127 (E)	20kV 15kV 10kV 380/220 (A) 220 (C)	± 5 (urban) ± 10 (rural)
IVORY COAST	50	220 (L) (1)	380/220 (A)	380/220 (A) (3)	(9)
JAMAICA	50 ± 1	220/110 (G (K)	220/110 (G) (K)	4/2.3kV 220/110 (G)	± 6
JAPAN (EAST) (4)	50 ± 0.2 (5)	200/100 (K) 100 (L)	200/100 (H) (K)	6.6kV 200/100 (H) 200 (G) (J)	± 10
JAPAN (WEST) (4)	60 ± 0.1 (5)	210/105 (K) 200/100 (K) 100 (L)	210/105 (H) (K) 200/100 (K) 100 (L)	22kV 6.6kV 210/105 (H) 200/100 (H)	± 10
JORDAN	50	380/220 (A) 220 (L)	380/220 (A)	380/220 (A) (3)	(9)
KENYA	50	240 (L) (1)	415/240 (A)	415/240 (A) (3)	(9)
KOREA	60	100 (L)	200/100 (K)	(9)	(9)
KUWAIT	50	240 (L) (1)	415/240 (A)	415/240 (A) (3)	(9)
LAOS	50 ± 8	380/220 (A)	380/220 (A)	380/220 (A) (3)	± 6
LEBANON	50	220 (L) (1) 110 (L) (1)	380/220 (A) 220 (L) 190/110 (A) 110 (L)	380/220 (A) (3) 190/110 (A)	(9)
LESOTHO	50	220 (L) (1)	380/220 (A)	380/220 (A) (3)	(9)
LIBERIA	60 ± 3.3	240/120 (K)	240/120 (K)	12.5/7.2kV 416/240 (B) 240/120 (K) 208/120 (D)	± 1.7
LIBYA	50	230 (L) (1) 127 (L) (1)	400/230 (A) 220/127 (A) 230 (L) 127 (L)	400/230 (A) (3) 220/127 (A)	(9)
LUXEMBOURG	50 ± 0.5	380/220 (A) 220 (L)	380/220 (A) 220 (L)	20kV 15kV	± 5 to 10
MALAGASSY REPUBLIC	50 ± 2	220 (L) (1) 127 (L) (1)	380/220 (A) 220/127 (A)	5kV 380/220 (A) 220/127 (A)	± 3
MALAWI	50	230 (L) (1)	400/230 (A)	400/230 (A) (3)	(9)
MALAYSIA	50 ± 1.0	240 (L) (1)	415/240 (A)	415/240 (A) (3)	+ 5 - 10
MALI	50	220 (L) (1) 127 (L) (1)	380/220 (A) 220/127 (A) 220 (L) 127 (L)	380/220 (A) (3) 220/127 (A)	(9)

World Electrical Supply Systems
Data — Courtesy of Canadian Standards Association

COUNTRY	FREQUENCY AND TOLERANCE Hz & %	HOUSEHOLD VOLTAGE V	COMMERCIAL VOLTAGE V	INDUSTRIAL VOLTAGE V	VOLTAGE TOLLERANCE V
MANILA	60 ± 5	240/120 (H) (K) 240/120 (H)	240/120 (H) (K) 240/120 (H)	20kV 6.24kV 3.6kV 240/120 (H)	± 5
MARTINIQUE	50	127 (L) (1)	220/127 (A) 127 (L)	220/127 (A) (3)	(9)
MAURITIUS	50 ± 1.0	230 (L) (1)	400/230 (A)	400/230 (A) (3)	± 6
MEXICO	60 ± 0.2	220/127 (A) 220 (L) 120 (M)	220/127 (A) 220 (L) 120 (M)	13.8kV 13.2kV 480/277 (A) 220/127 (B)	± 6
MONACO	50	380/220 (A) 220 (L) 220/127 (A) 127 (L)	380/220 (A) 220 (L)	380/220 (A) (3)	(9)
MONTSERRAT	60	230 (L) (1)	400/230 (A)	400/230 (A) (3)	(9)
MUSCAT & OMAN	50	240 (L) (1)	415/240 (A) 240 (L)	415/240 (A) (3)	(9)
MOROCCO	50	220/127 (A) 200/115 (A)	380/220 (A)	380/220 (A) (3)	(9)
NEPAL	50 ± 1	220 (L) (1)	400/220 (A) 220 (L)	11kV 400/220 (A)	± 10
NETHERLANDS	50 ± 0.4	380/220 (A) 220 (E) (L)	380/220 (A)	10kV 3kV 380/220 (A)	± 6
NETHERLANDS ANTILLES	50 & 60	220 (L) (1) 127 (L) (1) 120 (L) (1) 115 (L) (1)	380/220 (A) 230/115 (K) 220/127 (A) 208/120 (A)	380/220 (A) (3) 230/115 (G) 220/127 (A) 208/120 (A)	(9)
NEW GUINEA	50 ± 2	240 (L) (1)	415/240 (A) 240 (L)	22kV 11kV 415/240 (A)	± 5
NEW ZEALAND	50 ± 1.5	400/230 (A) (E) 230 (L) 240 (L)	415/240 (A) (E) 400/230 (A) (E) 230 (L)	11kV 400/230 (A) 415/240 (A) 440 (N) (6)	± 5
NICARAGUA	60	240/120 (G) (K)	240/120 (G) (K)	13.2kV 7.6kV 240/120 (G)	(9)
NIGERIA	50 ± 1	230 (L) (1) 220 (L) (1)	400/230 (A) 380/220 (A)	15kV 11kV 400/230 (A) 380/220 (A)	± 5
NIGER	50	220 (L) (1)	380/220 (A)	380/220 (A)	(9)
NORWAY	50 ± 0.2	230 (B)	380/220 (A) 230 (B)	20kV 10kV 5kV 380/220 (A) 230 (B)	± 10
PAKISTAN	50	230 (L) (1)	400/230 (A) 230 (L)	400/230 (A) (3)	(9)

World Electrical Supply Systems
Data — Courtesy of Canadian Standards Association

COUNTRY	FREQUENCY AND TOLERANCE Hz & %	HOUSEHOLD VOLTAGE V	COMMERCIAL VOLTAGE V	INDUSTRIAL VOLTAGE V	VOLTAGE TOLLERANCE V
PANAMA	60 ± 0.17	240/120 (K)	480/277 (A) 240/120 (K)	12kV 480/277 (A) 208/120 (A)	± 5
PARAGUAY	50	220 (L) (1)	440/220 (K) 380/220 (A)	440/220 (G) (3) 380/220 (A)	(9)
PERU	60	225 (B) (M)	225 (B) (M)	10kV 6kV 225 (B)	(9)
PHILIPPINES	60 ± 0.16	220/110 (K)	13.8kV 4.16kV 2.4kV 220/110 (H)	13.8kV 4.16kV 2.4kV 220/110 (H)	± 5
POLAND	50	220 (L) (1)	380/220 (A)	15kV 6kV 380/220 (A)	(9)
PORTUGAL	50 ± 5	380/220 (A) 220 (L)	15kV 5kV 380/220 (A) 220 (L)	15kV 5kV 380/220 (A)	± 1
PUERTO RICO	60 ± 10	240/120 (L)	480 (F) 240/120 (L)	8.32kV 4.16kV 480 (F)	± 10
QATAR	50	240 (L) (1)	415/240 (A) 240 (L)	415/240 (A) (3)	± 6
RHODESIA	50 ± 2.5	225 (L) (1)	390/225 (A)	11kV 390/225 (A)	± 6.6
ROMANIA	50 ± 1	220 (L) (1)	380/220 (L)	20kV 10kV 6kV 380/220 (A)	± 5
RWANDA	50 ± 1	220 (L) (1)	380/220 (A)	15kV 6.6kV 380/220 (A)	± 5
SABAH	50 ± 0.5	240 (L) (1)	415/240 (A)	415/240 (A) (3)	± 6
SAUDI ARABIA	60 ± 0.5	220/127 (A) 127 (L)	380/220 (A) 220/127 (A) 127 (L)	380/220 (A) (3) 220/127 (A)	± 5
SENEGAL	50	127 (L) (1)	220/127 (A) 127 (L)	220/127 (A) (3)	(9)
SIERRA LEONE	50	230 (L) (1)	400/230 (A) 230 (L)	11kV 400/230 (A)	(9)
SINGAPORE	50 ± 0.5	400/230 (A) 230 (L)	6.6kV 400/230 (A)	22kV 6.6kV 400/230 (A)	± 3
SOMALI REPUBLIC	50	230 (L) 220 (L) 110 (L) (1)	440/220 (K) 220/110 (K) 230 (L)	440/220 (G) (3) 220/110 (G)	(9)
SOUTH AFRICA	50 ± 2.5 25 (8)	433/250 (A) (7) 400/230 (A) (7) 380/220 (A) 220 (L)	11kV 6.6kV 3.3kV 433/250 (A) (7) 400/230 (A) (7) 380/220 (A)	11kV 6.6kV 3.3kV 500 (B) 380/220 (A)	± 5
SPAIN	50 ± 3	380/220 (A) (E) 220 (L) 220/127 (A) (E) 127 (L)	380/220 (A) 220/127 (A)	15kV 11kV 380/220 (A)	± 7

World Electrical Supply Systems
Data — Courtesy of Canadian Standards Association

COUNTRY	FREQUENCY AND TOLERANCE Hz & %	HOUSEHOLD VOLTAGE V	COMMERCIAL VOLTAGE V	INDUSTRIAL VOLTAGE V	VOLTAGE TOLLERANCE V
SRI LANKA CEYLON	50 ± 2	230 (L) (1)	400/230 (A) 230 (L)	11kV 400/230 (A)	± 6
ST. KITTS & NEVIS	60	230 (L) (1)	400/230 (A)	400/230 (A) (3)	(9)
ST. LUCIA	50	240 (L) (1)	415/240 (A)	11kV 415/240 (A)	(9)
ST. VINCENT	50	230 (L) (1)	400/230 (A)	3.3kV 400/230 (A)	(9)
SUDAN	50	240 (L) (1)	415/240 (A) 240 (L)	415/240 (A) (3)	(9)
SURINAM	50 & 60	115 (L) 127 (L) (1)	230/115 (K) 220/127 (A) 220/110 (K)	230/115 (G) (3) 220/127 (A) 220/110 (G)	(9)
SWAZILAND	50 ± 2.5	230 (L) (1)	400/230 (A) 230 (L)	11kV 400/230 (A)	± 6
SWEDEN	50 ± 0.2	380/220 (A) 220 (L)	380/220 (A) 220 (L)	20kV 10kV 6kV 380/220 (A)	± 10
SWITZERLAND	50 ± 0.5	380/220 (A) 220 (L)	380/220 (A) 220 (L)	16kV 11kV 6kV 380/220 (A)	± 10
SYRIA	50	220 (L) (1) 115 (L) (1)	380/220 (A) 220 (L) 200/115 (A) 115 (L)	380/220 (A) (3) 200/115 (A)	(9)
TAIWAN (FORMOSA)	60 ± 4	380/220 (A) 220 (L) 220/110 (K) 110 (L)	380/220 (A) 220/110 (H)	22.8kV 11.4kV 380/220 (A) 220 (H)	± 5 ± 10
TANZANIA	50	400/230 (A)	400/230 (A)	11kV 400/230 (A)	(9)
THAILAND	50	220 (L) (1)	380/220 (A) 220 (L)	380/220 (A) (3)	(9)
TOGO	50	220 (L) (1)	380/220 (A)	20kV 5.5kV 380/220 (A)	(9)
TONGA	50	415/240 (A) 240 (L) 110 (L)	415/240 (A) 240 (L) 110 (L)	11kV 6.6kV 415/240 (A)	(9)
TRINIDAD & TOBAGO	60 ± 0.5	230/115 (K)	400/230 (A) 230/115 (G)	12kV 400/230 (A)	± 6
TUNISIA	50 ± 2	380/220 (A) 220 (L)	380/220 (A) 220 (L)	15kV 1QkV 380/220 (A)	± 10
TURKEY	50 ± 2	220 (L) (1)	380/220 (A)	15kV 6.3kV 380/220 (A)	± 10
UGANDA	50 ± 0.01	240 (L) (1)	415/240 (A)	11kV 415/240 (A)	± 4.5
UNITED ARAB EMIRATES					
Dubai	50 ± 0.5	220 (L) (1)	380/220 (A) 220 (L)	6.6kV 380/220 (A)	± 2 to 3
Abu Dhabi	50	415/240 (A)	415/240 (A)	415/240 (A) (3)	(9)

World Electrical Supply Systems
Data — Courtesy of Canadian Standards Association

COUNTRY	FREQUENCY AND TOLERANCE Hz & %	HOUSEHOLD VOLTAGE V	COMMERCIAL VOLTAGE V	INDUSTRIAL VOLTAGE V	VOLTAGE TOLLERANCE V
UNITED KINGDOM (excluding Norhern Ireland)	50 ± 1	240 (L) (1)	415/240 (A)	22kV 11kV 6.6kV 3.3kV 415/240 (A)	± 6
URUGUAY	50 ± 1	220 (B) (L)	220 (B) (L)	15kV 6kV 220 (B)	± 6
USA (4)					
Charlotte (North Carolina)	60 ± 0.06	240/120 (K) 208/120 (A)	460/265 (A) 240/120 (K) 208/120 (A)	14.4kV 7.2kV 2.4kV 575 (F) 460 (F) 240 (F) 460/265 (A) 240/120 (K) 208/120 (A)	+ 5 - 2.5
Detroit (Michigan)	60 ± 0.2	240/120 (K) 208/120 (A)	480 (F) 240/120 (H) 208/120 (A)	13.2kV 4.8kV 4.16kV 480 (F) 240/120 (H) 208/120 (A)	+ 4
Los Angeles (California)	60 ± 0.2	240/120 (K)	4.8kV 240/120 (G)	4.8kV 240/120 (G)	± 5
Miami (Florida)	60 ± 0;.3	240/120 (K) 208/120 (A)	240/120 (K) 240/120 (H) 208/120 (A)	13.2kV 2;.4kV 480/277 (A) 240/120 (H)	± 5
New York (New York)	60	240/120 (K) 208/120 (A)	240/120 (K) 208/120 (A) 240 (F)	12.47kV 4.16kV 480/277 (A) 480 (F)	(9)
Pittsburgh (Pennsylvania)	60 ± 0.03	240/120 (K)	460/265 (A) 240/120 (K) 208/120 (A) 460 (F) 230 (F)	13.2 kV 11.5kV 2.4kV 460/265 (A) 208/120 (A) 460 (F) 230 (F)	± 5 (lighting) ± 10 (power)
Portland (Oregon)	60	240/120 (K)	480/277 (A) 240/120 (K) 208/120 (A) 480 (F) 240 (F)	19.9kV 12kV 7.2kV 2.4kV 480/277 (A) 208/120 (A) 480 (F) 240 (F)	(9)
San Francisco California	60 ± 0.08	240/120 (K)	480/277 (A) 240/120 (K)	20.8kV 12kV 4.16kV 480/277 (A) 240/120 (G)	± 5

World Electrical Supply Systems
Data — Courtesy of Canadian Standards Association

COUNTRY	FREQUENCY AND TOLERANCE Hz & %	HOUSEHOLD VOLTAGE V	COMMERCIAL VOLTAGE V	INDUSTRIAL VOLTAGE V	VOLTAGE TOLLERANCE V
Toledo (Ohio)	60 ± 0.08	240/120 (K) 208/120 (A)	480/277 (C) 240/120 (H) 208/120 (K)	12.47kV 7.2kV 4.8kV 4.16kV 480 (F) 480/277 (A) 208/120 (A)	± 5
RUSSIA	50	380/220 (A) 220 (L) 220/127 (A) 127 (L)	380/220 (A) 220 (L)	380/220 (A) (3)	(9)
VENEZUELA	60	240/120 (G) 208/120 (A)	240/120 (G) 208/120 (A)	13.8kV 12.47kV 4.8kV 4.16kV 2.4kV 240/120 (G) 208/120 (A)	(9)
VIETNAM	50 ± 0.1	220 (L) (1) 120 (L) (1)	380/220 (A) 208/120 (A)	15kV 380/220 (A)	± 10
YEMEN (PDR)	50 ± 1	250 (L) (1)	440/250 (A)	440/250 (A) (3)	± 4
YUGOSLAVIA	50	380/220 (A) 220 (L)	380/220 (A) 220 (L)	10kV 6.6kV 380/220 (A)	(9)
ZAÏRE, REPUBLIC OF	50	220 (L) (1)	380/220 (A)	380/220 (A) (3)	(9)
ZAMBIA	50 ± 2.5	230 (L) (1)	400/230 (A)	400/230 (A) (3)	± 3.75

NOTES:

1. The supply to each house is normally single-phase utilizing one phase line and neutral of systems (A) or (G).

2. Frequencies below 50Hz and d.c. supplies are in limited areas only. The supplies given indicates the diversity of possibilities which may exist.

3. Information on higher voltage supplies to factories not available.

4. More than one area of country has been given to illustrate the differences which exist. These may not be the only supplies available.

5. Frequency is 50 Hz (eastern area) and 60 Hz (western area). Dividing line is a North/South line through Shizuoka on Honshu Island.

6. Some areas are supplied via a single-wired earthed return (SWER) system (see Fig. N).

7. Only a few towns have this supply.

8. Refers to isolated mining districts.

9. Information not available at time of printing.

World Electrical Supply Systems
Data — Courtesy of Canadian Standards Association

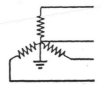

(A)
THREE-PHASE STAR;
FOUR-WIRE; EARTHED
NEUTRAL

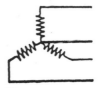

(D)
THREE-PHASE STAR;
FOUR-WIRE; NON-
EARTHED NEUTRAL

(G)
THREE-PHASE DELTA;
FOUR-WIRE; EARTHED
MID POINT OF PHASE

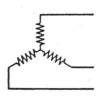

(B)
THREE-PHASE STAR;
THREE WIRE

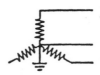

(E)
TWO-PHASE STAR;
THREE-WIRE; EARTHED
NEUTRAL

(H)
THREE PHASE OPEN
DELTA; FOUR-WIRE;
EARTHED MID POINT
OF PHASE

(C)
THREE PHASE STAR;
THREE-WIRE; EARTHED
NEUTRAL POINT

(F)
THREE PHASE DELTA;
THREE-WIRE

(J)
THREE PHASE OPEN;
DELTA; FOUR-WIRE;
EARTHED JUNCTION
OF PHASES

World Electrical Supply Systems

Data — Courtesy of Canadian Standards Association

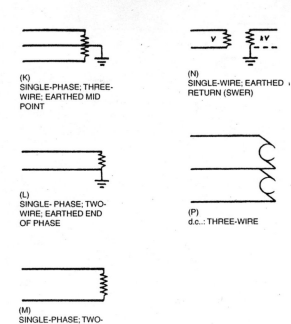

(K)
SINGLE-PHASE; THREE-
WIRE; EARTHED MID
POINT

(N)
SINGLE-WIRE; EARTHED
RETURN (SWER)

(L)
SINGLE- PHASE; TWO-
WIRE; EARTHED END
OF PHASE

(P)
d.c..: THREE-WIRE

(M)
SINGLE-PHASE; TWO-
WIRE; NON-EARTHED
NEUTRAL